UMTS - Signaling & Protocol Analysis

Volume 1: UTRAN and UE

INACON GmbH
Kriegsstrasse 154
76133 Karlsruhe
Germany
www.inacon.com
e-mail: inacon@inacon.de

Cover design by Stefan Kohler

© 1999 - 2005 INACON GmbH
Kriegsstrasse 154
76133 Karlsruhe

All rights reserved. No part of this publication may be reproduced, stored in a retrieval system, or transmitted by any means, electronic, mechanical, photocopying, recording, or otherwise, without written permission from the publisher. No patent liability is assumed with respect to the use of the information contained herein. Although every precaution has been taken in the preparation of this publication, the publisher and authors assume no responsibility for errors or omissions. Neither is any liability assumed for damages resulting from the use of the information contained herein. For more information, contact INACON GmbH at www.inacon.com.

UMTS - Signaling
&
Protocol Analysis

Volume 1: UTRAN and UE

Gunnar Heine

Legend:

All INACON publications use the same color codes to distinguish mandatory from optional or conditional parts in frame formats or optional from mandatory data blocks or signaling messages in scenarios. The different color codes are explained underneath:

- **Color Codes in Frame Formats:**

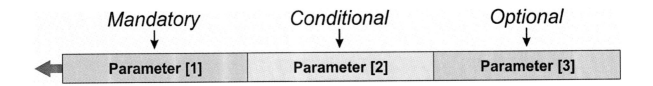

- **Color Codes in Scenarios:**

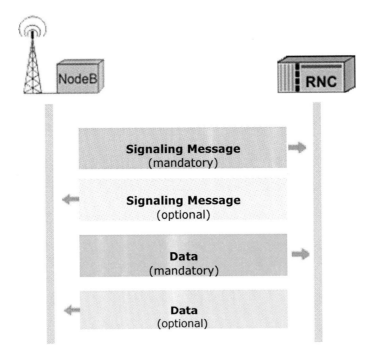

Foreword of the Publisher:

Dear Reader:

Note that this book is primarily a training document because the primary business of INACON GmbH is the training and consulting market for mobile communications. As such, we are proud to providing high-end training courses to many clients worldwide, among them operators like AT&T Wireless, INMARSAT or T-MOBILE and equipment suppliers like ERICSSON, MOTOROLA, NOKIA or SIEMENS.

INACON GmbH is not one of the old-fashioned publishers. With respect to time-to-market, form-factor, homogenous quality over all books and most importantly with respect to after-sales support, INACON GmbH is moving into a new direction. Therefore, INACON GmbH does not leave you alone with your issues and this book but we offer you to contact the author directly through e-mail (inacon@inacon.de), if you have any questions. All our authors are employees of INACON GmbH and all of them are proven experts in their area with usually many years of practical experience.

The most important assets and features of the book in front of you are:

- **Extreme degree of detailed information about a certain technology.**
- **Extensive and detailed index to allow instant access to information about virtually every parameter, timer and detail of this technology.**
- **Incorporation of several practical exercises.**
- **If applicable, incorporation of examples from our practical field experiences and real life recordings.**
- **References to the respective standards and recommendations on virtually every page.**

Finally, we again like to congratulate you to the purchase of this book and we like to wish you success in using it during your daily work.

Sincerely,

Gunnar Heine / President & CEO of INACON GmbH

PS: Please check for our UMTS online encyclopaedia at www.inacon.com.

UMTS – Signaling & Protocol Analysis

Table of Contents

■ A Comprehensive Inside View on UMTS 1

The UMTS Network Architecture 6
Release 99 6
Access Network (GERAN or UTRAN) 6
Circuit-Switched Core Network 6
Packet-Switched Core Network 6
The Different Types of RNC's 8
CRNC 8
SRNC 8
DRNC 8

The UMTS Protocol Stack with Rel. 99 10
Overview 10
The Circuit-Switched Control Plane 12
Access Stratum Protocols 12
Non Access Stratum Protocols 12
AAL-2 vs. AAL-5 12
The Circuit-Switched User Plane 14
The Packet-Switched Control Plane 16
Access Stratum Protocols 16
Non Access Stratum Protocols 16
The Packet-Switched User Plane 18
The Broadcast Domain (User Plane only) 20
Protocol Stack on Iur-Interface 22
The Transport Network Control Plane 24

The Physical Resource on the Uu-Interface (FDD) 26
Overview 26
Lower Parts of the Physical Layer 26
Upper Parts of the Physical Layer 26
Spreading and Modulation in Downlink Direction 28
Spreading and Modulation in Uplink Direction 30

An Overview of Channels in UMTS 32
Logical Channels 32
Transport Channels 32
Physical Channels 32

Logical-, Transport- and Physical Channels & the QoS 34
Example 34
Applications 34
Channels 34
Quality of Service 34

Details of Logical Channels 36
Control Channels 36
Traffic Channels 36

Details of Transport Channels 38
Common Transport Channels 38

UMTS – Signaling & Protocol Analysis

Dedicated Transport Channels 38
Transport Format and Transport Format Set... 40
Semi-Static Part.. 40
Dynamic Part... 40
Transport Block and Transport Block Set.. 42
Example: Transport Format Set Definition for PCH-TrCH 44
Example: Transport Format Set Definition for RACH-TrCH...................... 46

CCTrCH and Transport Format Combinations (UE-Side / FDD) 48
The CCTrCH... 48
Transport Format Combinations 48

Details of Physical Channels (FDD) 50
Dedicated Physical Channels 50
Common Physical Channels... 50
Layer 1 Signaling in Data Transferring Physical Channels 52
TPC-Field 52
Pilot Bits 52
FBI-Field... 52
TFCI-Field... 52
Example: Slot Format Definition for DPDCH/U and DPCCH/U................. 54
Example: Slot Format Definition for DPDCH/D and DPCCH/D................. 56
PCPCH-Data and Control (Physical Common Packet Channel) 58
PRACH-Data and Control (Physical Random Access Channel)............... 58
DPDCH/U (Dedicated Physical Data Channel / Uplink)............................ 58
DPCCH/U (Dedicated Physical Control Channel / Uplink)........................ 58
PDSCH (Physical Downlink Shared Channel).. 60
S-CCPCH (Secondary Common Control Physical Channel) 60
DPCCH/D for CPCH (Dedicated Physical Control Channel) 60
DPDCH/D and DPCCH/D (Downlink DPCH).. 60
PICH (Paging Indicator Channel).. 62
P-CCPCH (Primary Common Control Physical Channel)......................... 62
S-CPICH (Secondary Common Pilot Channel) 62
P-CPICH (Primary Common Pilot Channel) 62
S-SCH (Secondary Synchronization Channel) 62
P-SCH (Primary Synchronization Channel).. 62
CSICH (CPCH Status Indicator Channel) 64
CD/CA-ICH (CPCH Collision Detection / Channel Assignment Indicator Channel) 64
AP-AICH (CPCH Access Preamble Acquisition Indicator Channel) 64
AICH (Acquisition Indicator Channel)... 64

▪ Signaling & Protocol Analysis in UTRAN........... 65

Medium Access Control (MAC) 70
Overview: Tasks and Functions 70

Mapping of Logical, Transport and Physical Channels (FDD) 72
Mapping of BCCH and PCCH (UE-Perspective) 72
Mapping of DCCH and DTCH (UE-Perspective)...................................... 74
Mapping of CCCH (UE-Perspective).. 76
Mapping of CTCH (UE-Perspective).. 78

MAC-PDU Formats on Logical Channels (FDD / UE-Side) 80
Format on BCCH, PCCH and CTCH.. 80
Format on CCCH 82

UMTS – Signaling & Protocol Analysis

Format on DTCH and DCCH 84

The Different RNTI's 86
C-RNTI 86
DSCH-RNTI 86
S-RNTI 86
D-RNTI 86
U-RNTI 86
Example: Allocation of U-RNTI to UE... 88

The MAC-Architecture...92
MAC-b 94
MAC-c / sh 96
MAC-d 98

Traffic Volume Measurements and Radio Bearer Control 100

The Random Access Procedure 102
Initial Conditions 102
Overview 104
Access Service Class Selection 106
Transfer of an RRC_CONN_REQ-Message... 106
Other Cases (MLP) 106
Determination of the Persistence Value P 106
Persistence Value P and Random Number R... 108
PRACH Access Slots and AICH Access Slots 110
PRACH-Subchannels 110
Selection of an Appropriate PRACH-Access Slot..................................... 110
Response from the NodeB on AICH / Time Relation................................ 110
Preambles, Acquisition Indication and Message Part Transfer............... 112
RACH / PRACH Configuration in SIB 5 (Part 1).. 114
RACH / PRACH Configuration in SIB 5 (Part 2).. 116

Radio Link Control (RLC)...118
Operation Modes 118
Transparent Mode (TM) 118
Unacknowledged Mode (UM).. 118
Acknowledged Mode (AM) 118
Tasks and Functions.. 120
Configuration of the Size of an RLC-PDU 122

RLC-PDU-Types..124
The TMD-PDU 124
The UMD-PDU 124
Sequence Number (7 bit) 124
E-Bit... 124
Length Indicator (7 bit / 15 bit) plus E-Bit.. 124
Example of an RLC-TMD-PDU 126
Example of an RLC-UMD-PDU 128
TMD-PDU-Transfer (with segmentation) 130
TMD-PDU-Transfer (without segmentation) 132
UMD-PDU-Transfer... 134
The AMD-PDU 136
D/C-Bit... 136
Sequence Number (12 bit) 136
P-Bit... 136
HE-Field (2 bit) 136
Length Indicator (7 bit / 15 bit) plus E-Bit.. 136

UMTS – Signaling & Protocol Analysis

Example of an RLC-AMD-PDU	138
The STATUS- and Piggybacked STATUS-PDU (SUFI 1 – 4)	140
NO_MORE-SUFI	140
WINDOW-SUFI	140
ACK-SUFI	140
LIST-SUFI	140
Example of a STATUS-PDU with SUFI's LIST and ACK	142
The STATUS- and Piggybacked STATUS-PDU (SUFI 5 – 8)	144
BITMAP-SUFI	144
RLIST-SUFI	144
MRW-SUFI	144
MRW-ACK-SUFI	144
Example: Operation of the RLIST-SUFI	146
The RESET- and RESET-ACK-PDU	148

Overflow Protection ⇔ RLC-SDU-Discard **150**
- Timer Based RLC-SDU-Discard 150
  - With Signaling 150
  - Without Signaling 150
- Retransmission Based RLC-SDU-Discard 150
  - With Signaling 150
  - Reset-Procedure 150
- Transmit Buffer Overflow Based RLC-SDU Discard 150

Radio Resource Control (RRC) **152**
- Overview 152
- Tasks & Functions 154

Signaling Radio Bearers **156**
- Signaling Radio Bearer 0 (SRB 0) 156
- Signaling Radio Bearer 1 (SRB 1) 156
- Signaling Radio Bearer 2 (SRB 2) 156
- Signaling Radio Bearer 3 (SRB 3) 156
- Signaling Radio Bearer 4 (SRB 4) 156

The Different RRC-States **158**
- RRC-Idle Mode 160
  - UE is unknown in UTRAN 160
  - No DCCH's or DTCH's exist 160
  - UE monitors PICH / PCH in Downlink (⇔ DRX) 160
  - Change to RRC-Connected Mode requires Transmission of RRC_CONN_REQ ... 160
  - UE performs Autonomous Cell Reselection but neither Cell Updates nor URA Updates .. 160
  - UE Performs Routing and Location Area Update Procedures 160
- CELL_DCH-State 162
  - DCH's exist in Uplink and Downlink Direction 162
  - DCCH's are available and can be used; DTCH's may be available 162
  - UTRAN knows the Location of the UE on Cell Level 162
  - Handover Scenarios are Applicable 162
  - UE performs no Cell Updates or URA Updates 162
  - UE provides Measurement Reports to the RNC 162
- CELL_FACH-State 164
  - No DCH's exist in Uplink or Downlink Direction 164
  - DCCH's are available; DTCH's may be available 164
  - UE continuously monitors one FACH in Downlink 164
  - No Soft or Hard Handover Scenarios are applicable 164
  - UE performs Cell Updates but no URA Updates 164
  - UTRAN knows the Location of the UE on Cell Level 164
  - UE provides Measurement Reports to the RNC 164

UMTS – Signaling & Protocol Analysis

CELL_PCH-State 166
    No DCH's exist in Uplink and Downlink Direction 166
    DCCH's (and DTCH's) are configured but cannot be used in this State 166
    UE monitors PICH / PCH in Downlink (⇔ DRX) 166
    Uplink Transmission requires State Change to CELL_FACH (⇔ Cell Update) 166
    No Soft or Hard Handover Scenarios are applicable 166
    UE performs Cell Updates but no URA Updates 166
    UTRAN knows the Location of the UE on Cell Level 166
    UE provides Measurement Reports to the RNC 166
URA_PCH-State 168
    No DCH's exist in Uplink and Downlink Direction 168
    DCCH's (and DTCH's) are configured but cannot be used in this State 168
    UE monitors PICH / PCH in Downlink (⇔ DRX) 168
    Uplink Transmission requires State Change to CELL_FACH (⇔ Cell Update) 168
    No Handover Scenarios are Applicable 168
    UE performs URA Updates 168
    UTRAN knows the Location of the UE on URA Level 168
    UE provides Measurement Reports to the RNC 168
RRC State Transitions and Transitions to/from GSM 170

Encoding of RRC-Messages 172
ASN.1: Basic Encoding Rules (BER) 172
    Nested Information Elements 172
    ASN.1 Types 172
    The Sequence Type 172
    The Choice Type 172
    The Enumerated Type 172
    Value Assignment 172
TLV-Encoding 174
    The Tag-Field 174
    The Length Field 174
    Example: Encoding of IMSI 176
Extension of Tag- and Length-Fields 178
    Extension of the Tag-Value (Tag-Values > 30dec) 178
    Extension of the Length Value (Length Value > 127dec 178
    Undetermined Length 178
ASN.1: Packed Encoding Rules (PER) / Unaligned 180
    Erase Tag and Length Fields 180
    Minimize the Value Field 182
    Handling of Optional IE's 184
    Handling of IE's with Variable Length 186
Encoding Example: The RRC_CONN_REQ-Message 188

RRC-Message Types 192
Downlink DCCH-Messages 194
    ACT_SET_UPD 194
    ASSIST_DATA_DEL 194
    CELL_CHAN_UTRAN 194
    CELL_UPD_CNF 194
    COUNT_CHECK 194
    DL_DIR_TRANS 194
    HO_UTRAN_GSM 194
    HO_UTRAN_CDMA_2000 194
    MEAS_CTRL 196
    PAG_TYPE2 196
    PHYS_CHAN_RECONF 196
    PHYS_SHCH_ALL 196
    RB_RECONF 196

UMTS – Signaling & Protocol Analysis

RB_REL	196
RB_SETUP	196
RRC_CONN_REL	196
SEC_MODE_CMD	198
SIG_CONN_REL	198
TrCH_RECONF	198
TFC_CONTROL	198
UE_CAP_ENQ	198
UE_CAP_INF_CNF	198
UL_PHYS_CHAN_CTRL	198
URA_UPD_CNF	198
UTRAN_MOB_INFO	200

Uplink DCCH-Messages 202
- ACT_SET_UPD_COM 202
- ACT_SET_UPD_FAIL 202
- CELL_CHAN_UTRAN_FAIL 202
- COUNT_CHECK_RSP 202
- HO_UTRAN_COM 202
- INIT_DIR_TRANS 202
- HO_UTRAN_FAIL 202
- MEAS_CTRL_FAIL 202
- MEAS_REP 204
- PHYS_CHAN_RECONF_COM 204
- PHYS_CHAN_RECONF_FAIL 204
- RB_RECONF_COM 204
- RB_RECONF_FAIL 204
- RB_REL_COM 204
- RB_REL_FAIL 204
- RB_SETUP_COM 204
- RB_SETUP_FAIL 206
- RRC_CONN_REL_COM 206
- RRC_CONN_SETUP_COM 206
- RRC_STATUS 206
- SEC_MODE_COM 206
- SEC_MODE_FAIL 206
- SIG_CONN_REL_IND 206
- TrCH_RECONF_COM 206
- TrCH_RECONF_FAIL 208
- TFC_CTRL_FAIL 208
- UE_CAP_INFO 208
- UL_DIR_TRANS 208
- UTRAN_MOB_INFO_CNF 208
- UTRAN_MOB_INFO_FAIL 208

Downlink CCCH-Messages 210
- CELL_UPD_CNF 210
- RRC_CONN_REJ 210
- RRC_CONN_REL 210
- RRC_CONN_SETUP 210
- URA_UPD_CNF 210

Uplink CCCH-Messages 212
- CELL_UPD 212
- RRC_CONN_REQ 212
- URA_UPD 212

PCCH- (through PCH) and BCCH (through FACH)-Messages 214
- PAG_TYPE1 214
- SYS_INFO 214
- SYS_INFO_CHANGE_IND 214

UMTS – Signaling & Protocol Analysis

Measurements on the UE-Side	216
Example of a MEAS_CTRL-Message (Part 1)	218
Example of a MEAS_CTRL-Message (Part 2)	220
Important Measurement Parameters	222
UTRA Carrier RSSI	222
CPICH RSCP	222
CPICH Ec/No	222
Bit Error Rate on TrCH and Physical Channel	224
Example: QE in Iub-FP: UL-DATA represents the BER	226

▪ Important UMTS Scenarios & Call Tracing 227

Explanations of the Used Message Descriptors	**230**
Message Descriptors on Uu-Interface	230
Message Descriptors on Iu-cs-, Iu-ps- and Iur-Interface	230
Message Descriptors on Iub-Interface	232
NBAP-Messages (always over AAL-5)	232
RLC-PDU's (always over AAL-2)	232
ALCAP-Messages (over AAL-5)	232
Common Scenarios	**234**
NodeB Setup	234
Circuit-Switched Scenarios	**246**
UE-Registration	248
Initial Conditions	248
Applicability of this Procedure	248
Description	248
Mobile Originating Conversational Call	260
Initial Conditions	260
Applicability of this Procedure	260
Description	260
Mobile Terminating Conversational Call	284
Initial Conditions	284
Applicability of this Procedure	284
Description	284
Handover Scenarios	**308**
Variations of Soft Handover	310
Softer Handover (Radio Link Addition)	312
Initial Conditions	312
Applicability of this Procedure	312
Description	312
Softer Handover (Radio Link Removal)	316
Initial Conditions	316
Applicability of this Procedure	316
Description	316
Example: ACT_SET_UPD-Message (Link Removal)	318
Soft Handover (Intra-RNC / Inter-NodeB / Branch Addition)	320
Initial Conditions	320
Applicability of this Procedure	320
Description	320
Soft Handover (Inter-RNC / Branch-Addition)	326
Initial Conditions	326
Applicability of this Procedure	326
Description	326
Hard Handover UTRA-FDD $\Rightarrow$ GSM	334

UMTS – Signaling & Protocol Analysis

Overview	334
Hard Handover UTRA-FDD ⇒ GSM (Message Flow)	336
Initial Conditions	336
Applicability of this Procedure	336
Description	336
Example: Hard Handover UTRA FDD ⇒ GSM (Message Flow)	342
Hard Handover GSM ⇒ UTRA-FDD	346
Overview	346
Hard Handover GSM ⇒ UTRA-FDD (Message Flow)	348
Initial Conditions	348
Applicability of this Procedure	348
Description	348
UTRAN Mobility Management Procedures	**358**
Cell Update (Intra-RNC / Cell Reselection, Periodic, Page Rsp.)	360
Initial Conditions	360
Applicability of this Procedure	360
Description	360
Cell Update (Inter-RNC / Cell Reselection / UE initiated)	362
Initial Conditions	362
Applicability of this Procedure	362
Description	362
URA Update (Inter-RNC / Cell Reselection / UE initiated)	366
Initial Conditions	366
Applicability of this Procedure	366
Description	366
Packet-Switched Scenarios	**370**
Attachment	372
Initial Conditions	372
Applicability of this Procedure	372
Description	372
PDP-Context Activation (Mobile Originating)	384
Initial Conditions	384
Applicability of this Procedure	384
Description	384
Call Tracing in UTRAN	**398**
Introduction	398
Related Identifiers	398

- ## Enclosures for the Practical Exercises 403

- ## Solutions for the Practical Exercises 453

- ## References 458

- ## List of Acronyms 459

- ## Index: 477

UMTS – Signaling & Protocol Analysis

- **A Comprehensive Inside View on UMTS**
- **Signaling & Protocol Analysis in UTRAN**
- **Important UMTS Scenarios & Call Tracing**

UMTS – Signaling & Protocol Analysis

Intentionally left blank

Objectives

After this Lecture the Student will:

- **Understand the UMTS network architecture with Rel. 99 and the integration of UTRAN into the existing GERAN-network architecture**

- **Know in detail the different planes of the UMTS protocol stack**

- **Have an overview of the related UTRAN-protocols and their tasks and functions.**

- **Understand the channel concept of UMTS and how logical channels, transport channels and physical channels are mapped to each other.**

Objectives

- **Understand the UMTS network architecture with Rel. 99 and the integration of UTRAN into the existing GERAN-network architecture**
  UMTS reuses part of the existing GSM network architecture, namely the core network portion. Still, the understanding of the new UTRAN network architecture is essential to also understand the related signaling procedures.

- **Know in detail the different planes of the UMTS protocol stack**
  As a matter of fact, the UMTS protocol stack is that complex to require the split into various planes to allow for a complete comprehension. The term "various planes" relates namely to the user plane and the control plane as well as the transport plane and the transport network control plane.

- **Have an overview of the related UTRAN-protocols and their tasks and functions**
  In this first chapter there will only be an overview of the different UTRAN-protocols and their tasks and functions. Each protocol will be dealt with in detail in the related interface chapter.

- **Understand the channel concept of UMTS and how logical channels, transport channels and physical channels are mapped to each other**
  Although UMTS is highly flexible in the way how channels can be mapped to each other, there are still specific rules which prevent certain mappings. We have put together a complete overview on how the different channels and protocols fit together and are mapped together.

The UMTS Network Architecture

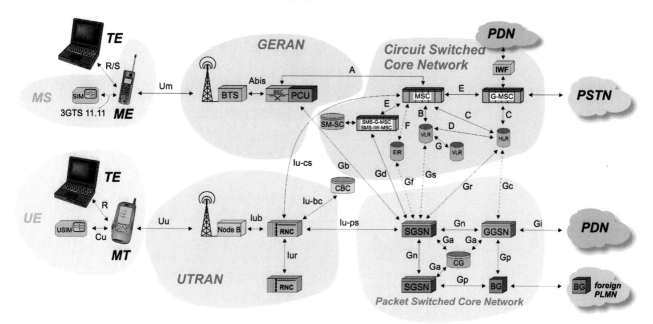

The UMTS Network Architecture

Release 99
The UMTS- and GSM-network architecture with Rel. 99 is illustrated in the figure. As the figure illustrates, the network architecture can be distinguished in:

Access Network (GERAN or UTRAN)
The GERAN is not scope of this document. It is fully dealt with in [1], [2] and [3]. The UTRAN is built from several RNS's (Radio Network Subsystem). Each RNS consists of exactly one RNC (Radio Network Controller) and several NodeB's. Between the RNC and each NodeB there is the Iub-interface while each RNC is connected to the MSC (Mobile Services Switching Center) through the Iu-cs-interface and to the SGSN (Serving GPRS Support Node) through the Iu-ps-interface.

Circuit-Switched Core Network
The circuit-switched core network consists most importantly of at least one MSC and at least one G-MSC (Gateway) and interconnects the mobile user to the PSTN (Public Switched Telephone Network). Integral part of the MSC is the VLR-database (Visitor Location Register), which provides for subscriber roaming functionality.

Packet-Switched Core Network
The packet-switched core network consists of at least one SGSN and at least one GGSN and interconnects the mobile user to one or more PDN's (Packet Data Network). To provide packet-switched services to roaming subscribers there is in addition the BG (Border Gateway). Finally, there is the CG (Charging Gateway) to store subscriber charging information.

> Note: When SGSN and MSC/VLR are integrated in a single physical unit they are called UMSC (UMTS-MSC).

For historical reasons, we illustrate the HLR (Home Location Register), the optional EIR (Equipment Identity Register) and the SM-SC (Short Message Service Center) to be part of the circuit-switched core network.

> Note: Many optional network elements have not been included like the entities for CAMEL, cell broadcast (CBC ⇔ Cell Broadcast Center) or for location based services (LCS).

[3GTS 23.002]

The Different Types of RNC's

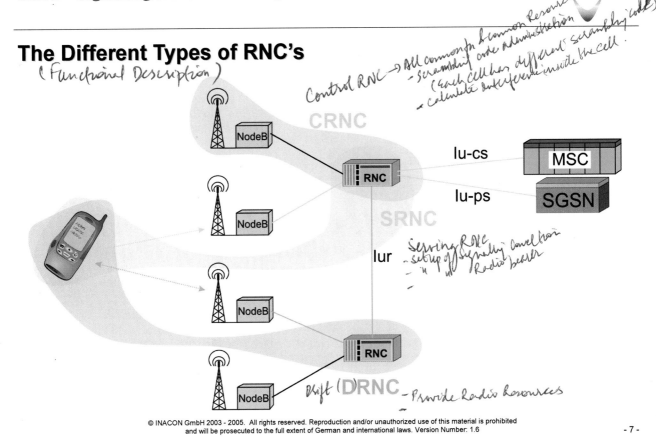

The Different Types of RNC's

CRNC
The CRNC or Controlling RNC "owns" the physical resources of each cell that is connected to it through the Iub-interface. The term CRNC has therefore a meaning exclusively between a given NodeB and an RNC. In our example, the upper RNC is the CRNC for the two NodeB's that are connected to it. We illustrate this through the "blue cloud".

SRNC
The SRNC or Serving RNC is defined as the very RNC that terminates the RRC-connection for a given UE and that interconnects the UE to the core network. The term SRNC is therefore applicable only between any UE with RRC-connection and exactly one RNC, the SRNC. In our example, the upper RNC is the SRNC for the UE. The "green cloud" and the green lines between UE, NodeB, SRNC and MSC / SGSN illustrate this.

DRNC
UMTS provides for the macrodiversity feature: One UE may receive data on physical channels from multiple NodeB's which may be connected to more than one RNC. The same applies vice versa in downlink direction. Only one of the possibly multiple involved RNC's is the SRNC that terminates the RRC-connection and which connects the UE to the core network. The other involved RNC's are referred to as DRNC's or Drift RNC's. In our example, the RNC in the bottom is a DRNC for the UE. The DRNC also provides a radio connection to the UE but the related information is exchanged through the Iur-interface with the SRNC.

[3GTS 25.401 (6)]

The UMTS Protocol Stack with Rel. 99

- Overview

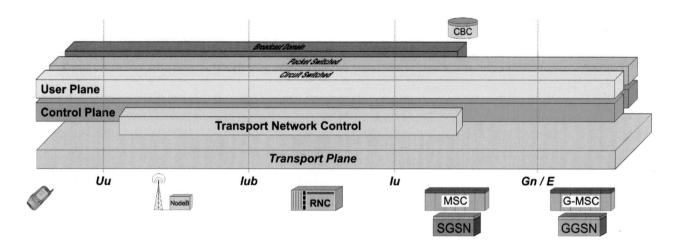

The UMTS Protocol Stack with Rel. 99

Overview
The UMTS protocol stack with Rel. 99 can roughly be split into 8 parts. The split is due to the separation of circuit-switched and packet-switched protocol stack on one hand and control plane and user (data) plane on the other hand. In addition to these consecutively 4 protocol stacks, there are more protocol stacks for the *broadcast domain*, for the protocols which are *applicable on the Iur-interface*, for the *transport network* and, last but not least, for the *transport network control plane*. Please note that the following more detailed presentations of the different UMTS protocol stacks will always include the currently defined transport network protocols (i.e. we won't show a separate protocol stack for the transport network).

- **The circuit-switched and packet-switched control-planes**
  The signaling protocols in the circuit-switched and packet-switched control planes are used to request, discuss, establish, maintain and release bearers for the transfer of user data. In addition, the control planes are used for mobility management functions.
- **The circuit-switched and packet-switched user (data) planes**
  The circuit-switched and packet-switched user planes are exclusively used to exchange user data of variable QoS and characteristics between the network and the user equipment.
- **The Broadcast Domain**
  The broadcast domain exists only in the user plane and provides for the transfer of broadcast information to UE's.
- **The transport network control plane**
  The transport network control plane consists of protocols which are *non-specific* to UMTS. As demanded by the UMTS-specific protocols like RANAP and NBAP, the transport network control plane protocols are responsible to allocate sufficient AAL-2 resources for the information exchange within UTRAN.

> Note that the transport network control plane is exclusively used to allocate AAL-2 resources! It is not required to allocate AAL-5 resources.

- **The Protocol Stack on the Iur-Interface**
  UMTS introduces the Iur-interface for the interconnection of two or more RNC's. This concept allows for macrodiversity functionality (soft handover).

> The separation of the UMTS-specific protocol stacks from the underlying transport network and its control plane shall provide more freedom and flexibility in the future to exchange or expand the transport network without affecting the UMTS core specifications.

[3GTS 25.401]

The Circuit-Switched Control Plane

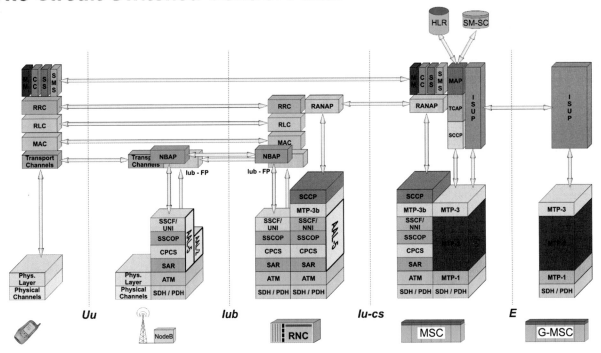

The Circuit-Switched Control Plane

The circuit-switched control plane is used for the exchange of control information which are related to circuit-switched services. In addition, the circuit-switched control plane is used for controlling supplementary services and it can be used for the exchange of short messages.

Access Stratum Protocols
The Access Stratum (AS) protocols are all protocols within the rectangle that is defined by the RRC-protocol (Radio Resource Control) at the upper left corner, the MAC-protocol (Medium Access Control) at the lower left corner and the RANAP-protocol (Radio Access Network Application Part) at the upper right corner.

Non Access Stratum Protocols
The Non Access Stratum (NAS) protocols could be all the remaining protocols but usually people refer to MM (Mobility Management), SS (Supplementary Services), CC (Call Control) and SMS (Short Message Services) as NAS-protocols within the circuit-switched control plane.

AAL-2 vs. AAL-5
Please note that AAL-5 is used for all control functions on the Iu-cs-interface (⇔ RANAP) and the Iub-interface (⇔ NBAP). On the other hand, the real-time AAL-2 is used for relaying UE-data and UE-signaling messages (⇔ Iub-FP) between NodeB and RNC and for user data on the Iu-cs-interface between RNC and MSC (next page).

[3GTS 25.412]

The Circuit-Switched User Plane

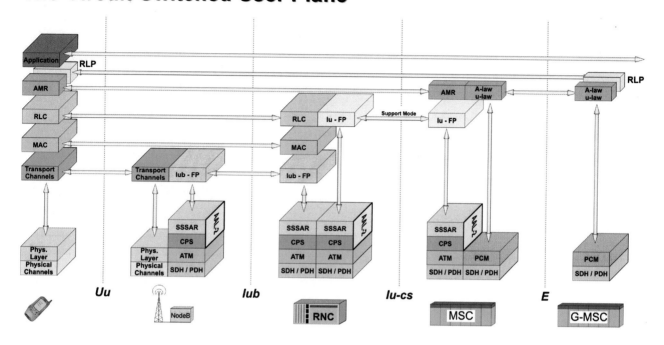

The Circuit-Switched User Plane

The circuit-switched control plane is exclusively used for the exchange of circuit-switched user data. In that respect, user data relates to standard voice transmission or multimedia or streaming applications.
The applications use AMR (Adaptive Multirate Encoding) for speech transmission and / or the RLP (Radio Link Protocol) for circuit-switched data transmission. Please recall that in UMTS multiple data streams can be configured, operated, re-configured and de-configured in parallel and simultaneously.
Please note that the Iu-FP is operated in "support Mode" within the circuit-switched user plane while it is operated in "transparent mode" in the packet-switched user plane.
The underlying transport network is using the real-time AAL-2 all the way on the Iub-interface and the Iu-cs-interface to provide for suitable QoS.

[3GTS 25.414]

The Packet-Switched Control Plane

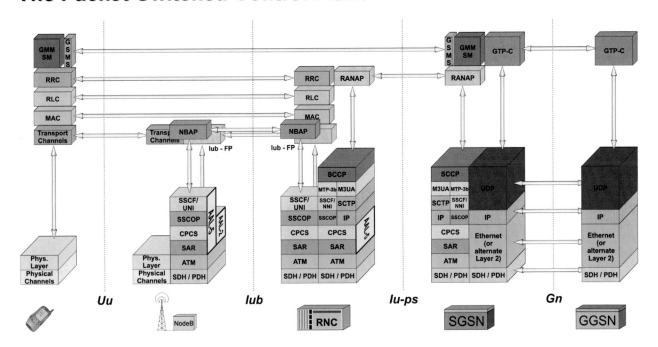

The Packet-Switched Control Plane

The packet-switched control plane is used for the exchange of control information which are related to packet-switched services. In addition, the packet-switched control plane can be used for the exchange of short messages.

Access Stratum Protocols
Like in the circuit-switched control plane, the Access Stratum (AS) protocols are all protocols within the rectangle that is defined by the RRC-protocol (Radio Resource Control) at the upper left corner, the MAC-protocol (Medium Access Control) at the lower left corner and the RANAP-protocol (Radio Access Network Application Part) at the upper right corner.

Non Access Stratum Protocols (All Beyond the occur area)
The Non Access Stratum (NAS) protocols are GMM (GPRS Mobility Management), SM (Session Management), and GSMS (SMS over packet-switched domain).

> Note that the Iu-ps-interface will always rely on ATM in the layer 1. However, as an implementation option the higher layers of the transport network may either use the broadband version of MTP-3 which is called MTP-3b or an IP-based network which uses the M3UA-protocol (MTP-3 User Adaptation) and the SCTP (Stream Control Transfer Protocol) to adjust IP to the SCCP (Signaling Connection Control Protocol).

[3GTS 25.412]

The Packet-Switched User Plane

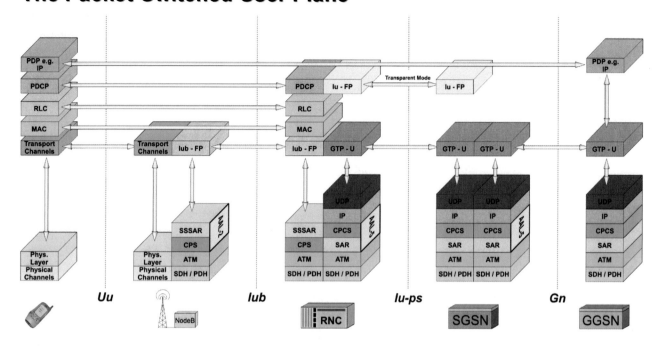

The Packet-Switched User Plane

The packet-switched user plane introduces two new protocols:
⇒ The PDCP (Packet Data Convergence Protocols) between the UE and the RNC for adjusting different types of packet data to the UMTS-protocol stack.
⇒ The GTP-U-protocol (GPRS Tunneling Protocol) between RNC and SGSN and between SGSN and GGSN for as bearer for packet data transfer.

Please note that GTP-U is relying on the non-real-time AAL-5 within the transport network.

[3GTS 25.414]

The Broadcast Domain (User Plane only)

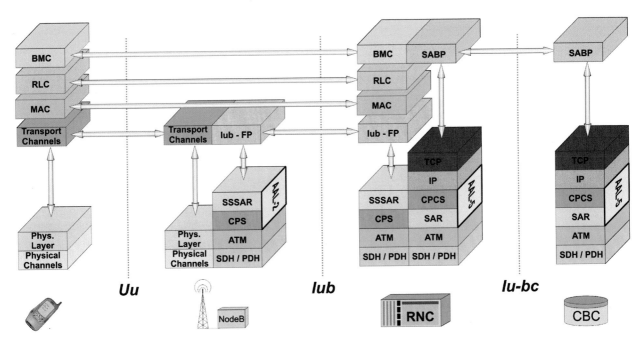

The Broadcast Domain (User Plane only)

The protocol stack in the broadcast domain is illustrated in the figure. It interconnects the CBC through the SABP-protocol (on Iu-bc-interface) and the BMC-protocol (on Iub-interface) towards the various UE's.

Note that SABP is using TCP/IP as path protocol.
Note that the broadcast domain is only defined within the user plane.

[3GTS 25.414]

Protocol Stack on Iur-Interface

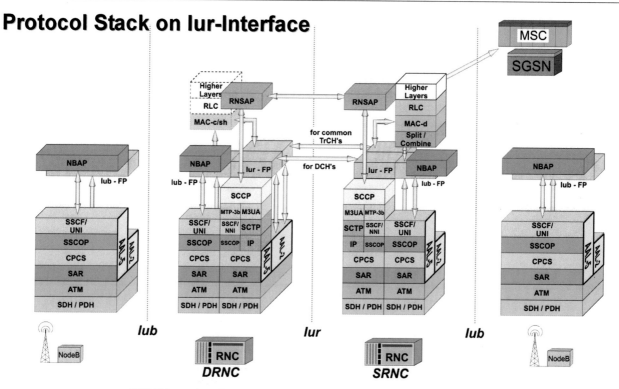

Protocol Stack on Iur-Interface

The Iur-interface is used to interconnect RNC's to provide for macrodiversity (⇔ soft handover) and a faster information exchange between RNC's in general. The figure illustrates how a DRNC (Drift-RNC) interconnects a NodeB through the Iur-interface towards the SRNC (Serving RNC).
The RNSAP-protocol is used to control the establishment, maintenance and release of interconnections through the Iur-interface. These interconnections are realized by AAL-2 virtual connections for both packet-switched and circuit-switched traffic. The RNSAP-protocol is used in addition to relay information on CCCH between the SRNC and the DRNC.

As the figure illustrates, there are two possibilities how to transfer DCCH- / DTCH-data over the Iur-interface using AAL-2:

1. If dedicated TrCH's (DCH's) are used on the Uu-interface, an AAL-2-link is established between DRNC and SRNC which is used by the Iur-FP to relay the DCH-transport block sets over the Iur-interface. The information is split / combined in the SRNC.
2. If common TrCH's are used on the Uu-interface, there is also an AAL-2-link established between DRNC and SRNC to relay the DCCH- and/or DTCH-information between the SRNC and the DRNC over common TrCH data streams (Iur-FP for common TrCH's). Please note in the figure how this link allows the SRNC to use the MAC-c/sh resources in the DRNC.

Note:
- Only in case 1. (⇔ DCH's are used) the SRNC features the specific split/combine-function between Iub-FP and MAC.
- Therefore, soft handover and macrodiversity can only be used, if DCH's are equipped.

[3GTS 25.422, 3GTS 25.424, 3GTS 25.425, 3GTS 25.426]

The Transport Network Control Plane

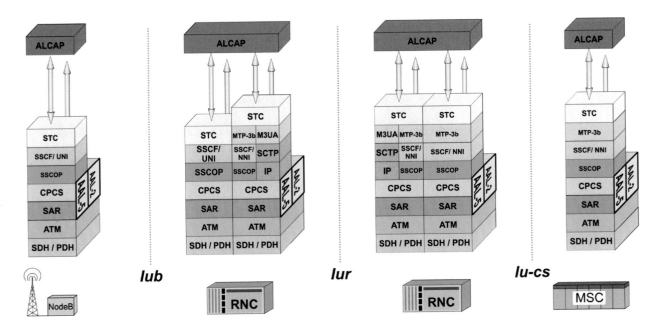

The Transport Network Control Plane

The transport network control plane introduces yet another protocol stack which is illustrated in the figure. This transport network control plane ranks from the Node B towards the MSC.
At the core of the transport network control plane is the ALCAP-protocol (Access Link Control Application Part). You will sometimes also find the synonymous term AAL2L3-protocol.
ALCAP resides on AAL-5 as bearer but it is exclusively used to control AAL-2 connections between two adjacent network elements. In that respect, ALCAP is only a slave of the NBAP-protocol on Iub-interface, of the RANAP-protocol on Iu-cs-interface and of the RNSAP-protocol on Iur-interface. If triggered by these protocols, ALCAP has to react with the provision, addition, reconfiguration or release of the respective AAL-2-resources.

[3GTS 25.414, 3GTS 25.426]

The Physical Resource on the Uu-Interface (FDD)

- Overview

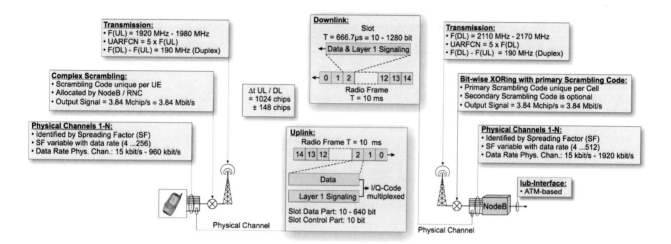

The Physical Resource on the Uu-Interface (FDD)

Overview

Lower Parts of the Physical Layer

⇒ The information flow between UE and NodeB is based on W-CDMA as access scheme. Although both operation modes, TDD and FDD, are defined for UMTS, as of today only FDD is used.

⇒ Accordingly, UE and NodeB operate on paired band frequencies with a duplex distance of 190 MHz. Each UMTS-frequency channel is 5 MHz wide and identified through the UARFCN. In addition to the primary band illustrated in the figure, there is also a second frequency band defined for UTRA-FDD (DL: 1930 MHz – 1990 MHz; UL: 1850 MHz – 1910 MHz; Duplex Distance: 80 MHz).

⇒ Usually, all NodeB's within a PLMN transmit on the *same* downlink frequency at the *same* time. Vice versa, all UE's transmit on the *same* uplink frequency possibly at the same time.

⇒ The distinction among the various NodeB's in downlink direction and the various UE's in uplink direction is accomplished through a unique scrambling code. Scrambling codes are non-orthogonal pseudo-noise codes with good auto-correlation characteristics [4]. The length is either 256 chips (⇔ short codes) or 38400 chips (⇔ long codes / Gold-tree). In uplink direction, several million scrambling codes of short or long length can be used of which every cell of NodeB's in a cluster is using only a subset (to provide for macrodiversity). In downlink direction, only 512 different scrambling with long length (38400 chips) are applicable. Hence, the UE can theoretically distinguish among up to 512 different cells.

Upper Parts of the Physical Layer

⇒ The upper parts of the physical layer are dominated by the physical channels. Different physical channel types exist for different tasks. Physical channels can have data rates (after channel coding) of 15 kbit/s – 1920 kbit/s in downlink direction and 15 kbit/s – 960 kbit/s in uplink direction. The lower maximum data rate in uplink direction is due to the I/Q-code multiplexing.

⇒ The data rates depend on the variable length spreading factor. In downlink direction, spreading factors are 4 bit – 512 bit long, in uplink direction, the length is 4 bit – 256 bit. Spreading factors are fully orthogonal codes with good cross-correlation characteristics to distinguish the different physical channels of a single source from each other. The longer the spreading factor, the lower the data rate. Spreading factors are arranged in code trees and are therefore not independent from each other. In other words: Theoretically, in a given cell in downlink direction there could be only 4 users with a spreading factor of 4 but 512 users with a spreading factor of 512. In practice, some spreading factor codes are reserved for administrative channels and can therefore not be allocated for user data transfer.

⇒ In the time domain, physical channels are organized in radio frames with a duration of 10 ms. Each radio frame is split into 15 slots with a duration of 666.67 µs. The format of the single slots varies with the physical channel type and is controlled by the RNC.

⇒ Uplink and downlink transmission are time aligned. The downlink transmission shall start 1024 ±148 chips before uplink transmission.

Spreading and Modulation in Downlink Direction

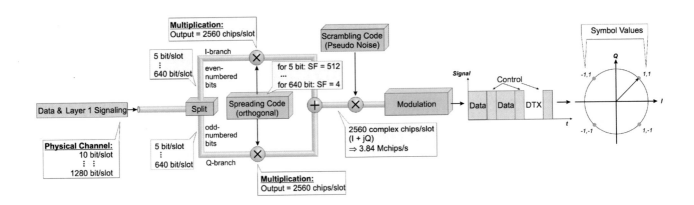

Spreading and Modulation in Downlink Direction

⇒ In downlink direction, the data and layer 1 signaling part on a single physical channel are time multiplexed within the 15 slots of a single radio frame (10 ms). Several different types of physical channels are defined. Depending on the type of a physical channel and depending on the spreading factor (SF = 4 ... 512), the combined number of data and control bits per slot varies between 10 bit and 1280 bit.
⇒ In a first step, the bit stream is split into two parts: The even-numbered bits are considered as the I-branch while the odd-numbered bits are considered as the Q-branch. The Q-branch will be +90°-phase shifted to the I-branch.
⇒ In the next step, both branches are multiplied by the same orthogonal spreading code (SF = 4 ... 512) which is also called channelization code. The output bit stream will therefore always contain 2560 bit.

Note:
- For a better differentiation from the unspreaded bit stream, the content of the output bit stream of the spreading unit is referred to as chips.
- In downlink direction, the different physical channels are distinguished through different channelization codes. Consequentially, a given UE will only listen to those physical channels that it has been allocated.
- The channelization codes which are used in UTRA are taken from the Walsh-tree (OVSF-codes). Between 4 and 512 physical channels can be distinguished through these codes.

⇒ After the channelization, the two branches (I + Q) are de-multiplexed and form together a single complex data stream of (I + jQ)-symbols.
⇒ These complex symbols are scrambled with 1 out of 512 complex scrambling codes (⇔ pseudo-noise code) to uniquely identify the physical signals from a single NodeB. Since the scrambling code has the same chip rate as the already spread signal, it does not change the chip rate. The final step in the processing chain is the QPSK-modulation.
⇒ The figure also illustrates the Signal / Time-diagram to emphasize the impact of DTX (Discontinuous Transmission) on the signal characteristics. During DTX-periods, the output signal will be entirely switched off to reduce the downlink interference.
⇒ Please note the different symbol values in the I/Q-diagram that distinguish between (+1) and (−1). Actually, the conversion of bit values (1) and (0) into (+1) and (-1) is conducted to provide for an easier multiplication during channelization.

Note:
In UMTS, bit values are converted the following way:
- **Real:** Bit Value 1 ⇒ (−1), Bit Value 0 ⇒ (+1)
- **Complex:** Bit Sequence '00' ⇒ (+1 + j)   Bit Sequence '01' ⇒ (−1 + j)
  Bit Sequence '10' ⇒ (+1 − j)   Bit Sequence '11' ⇒ (−1 − j)

Spreading and Modulation in Uplink Direction

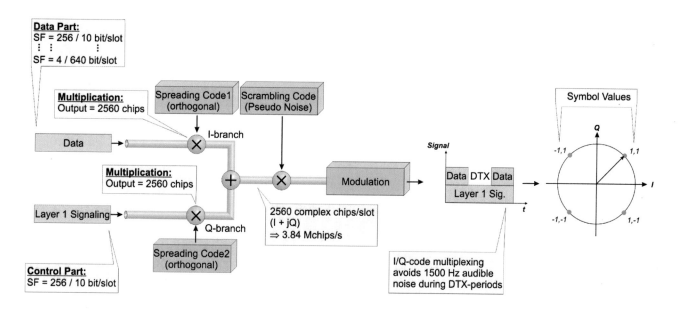

Spreading and Modulation in Uplink Direction

⇒ The figure illustrates the format of physical channels in uplink direction. Please consider that in uplink direction, the data and control part are processed and transmitted in parallel.
⇒ The data part of each slot within a radio frame (10 ms) will contain between 10 and 640 channel coded and interleaved data bits depending on the spreading factor (SF = 4 … 256). The control part (⇔ layer 1 signaling) will always contain 10 bit and the respective spreading factor will always be 256.

Note that the two spreading factors for the data and control part are different from each other.

⇒ The data part is considered the I-branch of the processing chain while the control part forms the Q-branch. As in downlink direction, the spreading process is also called channelization process. And like in downlink direction, perfectly orthogonal spreading codes out of the Walsh-tree are used. The output of the channelization process always contains 2560 chips/slot.

Note:
- Since I/Q-Code Multiplexing is used in uplink direction, the data part and the control part share the resources 50:50. Since both the I- branch and the Q-branch feed minimum 10 bit into the channelization process, the minimum channelization code in uplink direction is SF = 256 (unlike SF = 512 in downlink).
- The maximum gross data rate is only half of what it could be without I/Q-code multiplexing, because half of the de-multiplexed chip stream belongs to the control part.

⇒ After the channelization, the two branches (I + Q) are de-multiplexed and form together a single complex data stream of (I + jQ)-symbols.
⇒ These complex symbols are scrambled with the complex scrambling code (⇔ pseudo-noise code) to uniquely identify the physical signals from one UE. In uplink direction, several million scrambling codes (Long Code No = 0 … 16.777.215 (⇔ 38400 chips) or Short Code Number = 0 … 16.777.215 (⇔ 256 chips) are defined. The very scrambling code to be used by the UE is allocated by the RNC.
⇒ The final step in the processing chain is the QPSK-modulation.

The figure also illustrates the Signal / Time-diagram to emphasize the impact of DTX (Discontinuous Transmission) on the signal characteristics. During DTX-periods, the output signal will not be entirely switched off to avoid a 1500 Hz audio pattern in close-by audio equipment.

An Overview of Channels in UMTS

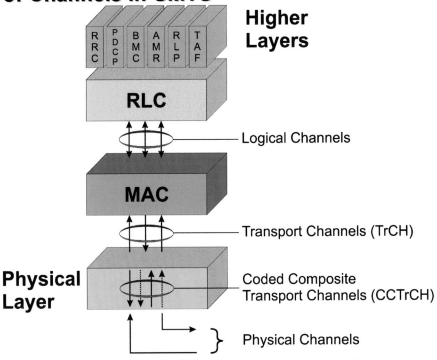

An Overview of Channels in UMTS

> Note: The comprehension of the channel concept, the channel mapping and the functions and differences of the different channels is *the* milestone in the understanding how UMTS operates.

In general, there are 3 different types of channels defined for the UMTS:

Logical Channels
Logical channels provide the bearers for the information exchange between the MAC-protocol and the RLC-protocol. Logical channels can be unidirectional or bi-directional. There are two types of logical channels:
⇒ Traffic channels to convey user data within the user data planes.
⇒ Control channels to convey signaling information within the control planes.

Transport Channels
Transport Channels provide the bearers for the information exchange between the MAC-protocol and the physical layer. All transport channels are unidirectional.
Within the physical layer there is multiplexing unit that maps the transport channels of one type (e.g. all DCH's) onto one CCTrCH. The data from one CCTrCH is finally mapped onto one or more physical channels. For each UE, there can be only one CCTrCH per transport channel type.

Physical Channels
Physical channels provide the bearers for the different CCTrCH's / TrCH's. Each physical channel is identified through its frequency (UARFCN), spreading code, scrambling code and duration. In uplink direction, the phase of the signal (0 or 90°) is also required for identification and differentiation.

[3GTS 25.301, 3GTS 25.302, 3GTS 25.211]

Logical-, Transport- and Physical Channels & the QoS

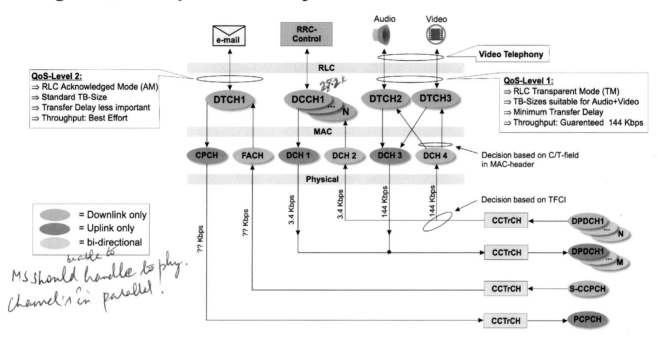

Logical-, Transport- and Physical Channels & the QoS

Example
⇒ The figure illustrates how the channel concept in UMTS provides for suitable QoS for different simultaneous applications (UE-perspective).

Note that the presented case will only work, if the UE supports simultaneous reception on one or more DPDCH's plus on S-CCPCH for the FACH-TrCH. This is an optional feature.

Applications
⇒ In the presented case, the UE-user is simultaneously having a video telephony call with different data streams for video and audio signals plus e-mails are uploaded and/or downloaded.
⇒ Obviously, an RRC-connection exists which is used to exchange RRC-control information between the RNC and the UE.

Channels
⇒ The video and audio data streams are mapped on dedicated logical traffic channels (DTCH 2 and DTCH 3). DTCH 2 and 3 are mapped on the same dedicated TrCH's in uplink (DCH 3) and downlink (DCH 4) direction. Through the C/T-field, the incoming and outgoing MAC-PDU's are assigned either to DTCH 2 or DTCH 3.
⇒ The different DCCH's for the different signaling radio bearers are also mapped on DCH's (DCH 1 and 2) but these DCH's offer much less throughput rate (e.g. 3.4 kbit/s).

The throughput rate of a TrCH is controlled through the transport block set size and the TTI (Transmission Time Interval).

⇒ In each direction, the DCH's are de-multiplexed on only one CCTrCH and then on one or more DPDCH. Obviously, in each direction there is also an associated DPCCH for the one or more DPDCH's. The TFCI-field in the DPCCH is used to distinguish content for the different DCH's.
⇒ The e-mail transfer uses DTCH 1 which is mapped on common TrCH's. In uplink direction, a CPCH is used while in downlink direction a FACH is used. Accordingly, the UE requires a C-RNTI to be able to use the CPCH. The RNC also uses the C-RNTI to address the UE on FACH.

Quality of Service
⇒ Obviously, the QoS-profile for the two different applications is determined beforehand. For the e-mail transfer we selected acknowledged RLC-mode for secure transmission which however implies a less important transfer delay. The throughput rate on the common TrCH's cannot be determined since it is a shared resource. However, the RNC may guarantee a certain bitrate to one user and prioritize that user also on the common TrCH's at least in downlink direction. In uplink direction the throughput rate on common TrCH's cannot be guaranteed.
⇒ For the video telephony the QoS-requirements are vice versa. The DCH's as TrCH's provide a guaranteed low throughput rate, if deployed together with RLC transparent mode.

Details of Logical Channels

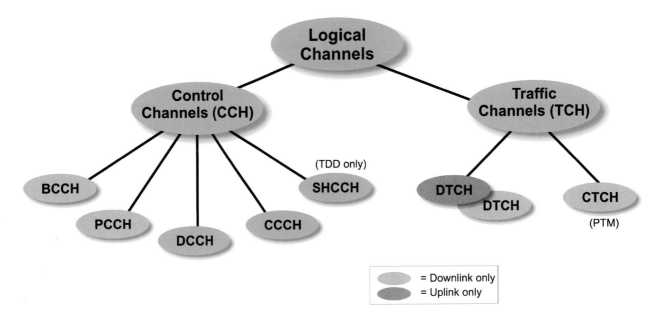

Details of Logical Channels

Control Channels
Control channels are exclusively used in the control plane.

- **BCCH (Broadcast Control Channel)**
  The BCCH is usually mapped on the BCH-TrCH and it is used to broadcast RRC-system information blocks to the UE's. Only SIB 10 (System Information Block) can also be mapped on FACH-TrCH.

- **PCCH (Paging Control Channel)**
  The PCCH is exclusively mapped on the PCH-TrCH and it is used to convey RRC: PAG_TYPE1-messages to UE's in RRC_IDLE-mode, CELL_PCH-state and URA_PCH-state.

- **DCCH (Dedicated Control Channel)**
  The DCCH can be mapped on different TrCH's (depending on the RRC-connection state) and it is used to provide a bearer for signaling radio bearers SRB 1, SRB 2, SRB 3, SRB 4, ... The presence of the DCCH requires that an RRC-connection exists. However, DCCH's can only be used in CELL_FACH- and CELL_DCH-state.

- **CCCH (Common Control Channel)**
  The CCCH is used to convey messages on SRB 0 between UE ad CRNC. In downlink direction, the CCCH will always be mapped on the FACH-TrCH while in uplink direction, the CCCH is mapped on the RACH-TrCH. The CCCH is used to establish an RRC-connection and it is used to perform UTRAN-mobility management procedures (cell update, URA-update).

- **SHCCH (Shared Channel Control Channel)**
  TDD only

Traffic Channels
Traffic channels are exclusively used in the user (data) plane.

- **DTCH (Dedicated Traffic Channel)**
  The DTCH is used to convey user plane data between the UE and the RNC. DTCH's can use common or dedicated TrCH's (depending on QoS-requirements) and are used as well for the transfer of voice call data as for the transfer of e-mails.

- **CTCH (Common Traffic Channel)**
  The CTCH is a unidirectional point-to-multipoint traffic channel that is exclusively used by the CBC (through the RNC) to convey data to UE's.

[3GTS 25.301 (5.3.1.1.1)]

Details of Transport Channels

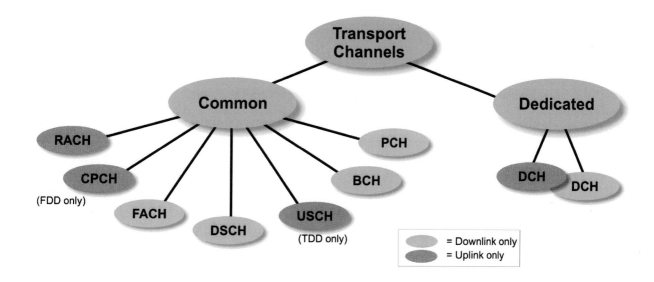

Details of Transport Channels

The different transport channels differ in their transport characteristics which is configured in the transport format or transport format set (⇔ consisting of more than one transport format). A transport format defines how the physical layer shall process an information block on a transport channel with respect to channel coding, interleaving etc.. Transport Channels are distinguished into:

Common Transport Channels
Unlike dedicated transport channels, common transport channels require an inband identification of the target or source UE (through the RNTI). Except for the BCH which is controlled by the NodeB, all common transport channels are controlled by the CRNC. The following common transport channels are defined:

- **RACH (Random Access Channel)**
- **FACH (Forward Access Channel)**
- **USCH (Uplink Shared Channel)**
- **PCH (Paging Channel)**
- **CPCH (Common Packet Channel)**
- **DSCH (Downlink Shared Channel)**
- **BCH (Broadcast Channel)**

Dedicated Transport Channels
On dedicated transport channels, the UE is identified through the spreading factor code (⇔ downlink) or the scrambling code (⇔ uplink), respectively. Dedicated transport channels are always controlled by the SRNC. There is only one type of dedicated transport channel defined:

- **DCH (Dedicated Channel)**

In addition, there is the CCTrCH (Coded Composite Transport Channel) that combines all transport channels of the same type within the physical layer.

[3GTS 25.302 (6/7), 3GTS 25.211 (4.1)]

Transport Format and Transport Format Set

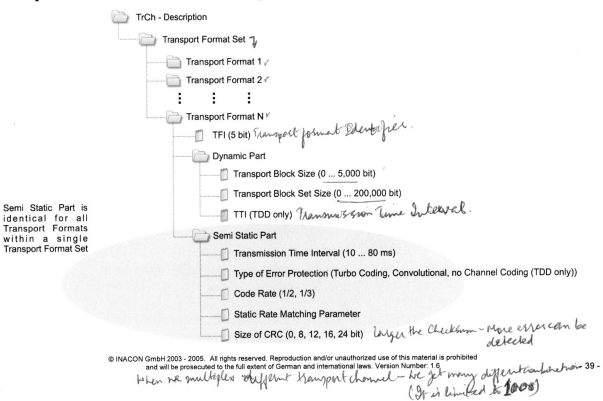

Transport Format and Transport Format Set

The characteristics of a given transport channel is highly flexible. Upon activation of a transport channel, this characteristics is configured by the RNC through the RRC-protocol. It can be reconfigured by the RRC-protocol at any time. As mentioned before, this characteristics is referred to as transport format (TF). For a given transport channel, more than one transport format can be configured. Each transport format is identified between MAC and physical layer through the TFI (Transport Format Indicator / 5 bit / TFI = 0 … 31). This TFI can be viewed on Iub-interface in the UL-DATA- and DL-DATA-PDU's. The whole set of transport formats that is applicable on one configured transport channel is referred to as transport format set (TFS).
Each transport format set consists of the following parts:

Semi-Static Part
The semi-static part is identical for all transport formats of a given transport channel. It consists of the:
⇒ TTI (Transmission Time Interval / 10 – 80 ms) that defines the time interval between two consecutive deliveries of transport block sets for the same transport channel to the physical layer.

Note: If real-time services require a very low transfer delay, a TTI of 80 ms is inappropriate because the oldest TB per TBS will already be 80 ms old when the TBS is delivered.

⇒ Type of error protection to be used for a transport block set.
⇒ Channel code rate (1/2, 1/3)
⇒ Rate matching parameter that tells the physical unit whether to puncture or to pad the channel coded transport block set to the required size. The required size is given by the number of data bit positions on the physical channel.
⇒ Size of the CRC-field.

Dynamic Part
Unlike the semi-static part, the dynamic part differs among the transport formats of a transport format set. The dynamic part consists of the:
⇒ Transport block size (in bit)
⇒ Transport block set size (in bit)
⇒ and optionally (in TDD only) the TTI.

Note: All transport block within one transport block set shall have the same size. Accordingly: No_of_TB's_per_TBS = TBS-Size / TB-Size

[3GTS 25.302 (7.1)]

Transport Blocks and Transport Block Sets

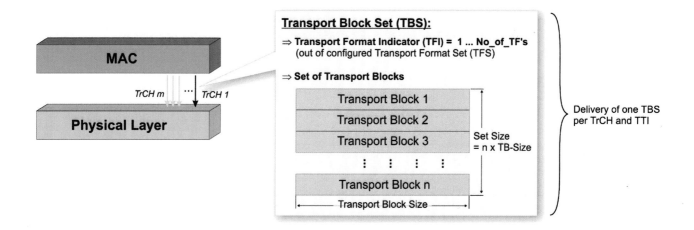

Transport Block and Transport Block Set

The figure illustrates the delivery of a transport block set from MAC to the physical layer on TrCH 1. Obviously, there may be more TrCH's 2 – n configured between MAC and the physical layer for which different transport format sets are applicable.

Note:
- The primitive which delivers the TBS to the physical layer will also contain the TFI to tell the physical layer up-front, how many transport blocks of which size are included.
- The channel coding scheme etc. are already known by the physical layer for each TrCH as they belong to the semi-static part of the TFS.

[3GTS 25.302 (7.1.1) / (7.1.2)]

Example: Transport Format Set Definition for PCH-TrCH

RNC & NodeB messaging

```
+---------+------------------------------------------+------------------------------------+
|         |1.5.1.3.3.1.9 pCH-Parameters              |                                    |
|***B2*** |1.5.1.3.3.1.9.1 id                        |id-PCH-ParametersItem-CTCH-SetupRqst|
|00------ |1.5.1.3.3.1.9.2 criticality               |reject                              |
|         |1.5.1.3.3.1.9.3 value                     |                                    |
|00001100 |1.5.1.3.3.1.9.3.1 commonTransportChannelID|5                                   |
|         |1.5.1.3.3.1.9.3.2 transportFormatSet      |                                    |
|         |1.5.1.3.3.1.9.3.2.1 dynamicParts          |                                    |
|         |1.5.1.3.3.1.9.3.2.1.1 sequence            |                                    |
|***B2*** |1.5.1.3.3.1.9.3.2.1.1.1 nrOfTransportBlocks|0                                  |
|         |1.5.1.3.3.1.9.3.2.1.1.2 mode              |                                    |
|         |  1.5.1.3.3.1.9.3.2.1.1.2.1 notApplicable |0                                   |
|         |1.5.1.3.3.1.9.3.2.1.2 sequence            |                                    |
|***B2*** |1.5.1.3.3.1.9.3.2.1.2.1 nrOfTransportBlocks|1                                  |
|***B2*** |1.5.1.3.3.1.9.3.2.1.2.2 transportBlockSize|240                                 |
|         |1.5.1.3.3.1.9.3.2.1.2.3 mode              |                                    |
|         |  1.5.1.3.3.1.9.3.2.1.2.3.1 notApplicable |0                                   |
|         |1.5.1.3.3.1.9.3.2.2 semi-staticPart       |                                    |
|***b3*** |1.5.1.3.3.1.9.3.2.2.1 transmissionTimeInter..|msec-10  Transmission Time Interval|
|--01---- |1.5.1.3.3.1.9.3.2.2.2 channelCoding       |convolutional-coding  Two conv. codecs used - 1602 & 1603|
|-----0-- |1.5.1.3.3.1.9.3.2.2.3 codingRate          |half                                |
|11100101 |1.5.1.3.3.1.9.3.2.2.4 rateMatchingAttribute|230                                |
|-011---- |1.5.1.3.3.1.9.3.2.2.5 cRC-Size   → 16 bits|v16                                 |
+---------+------------------------------------------+------------------------------------+
```

⇒ Only 1 transport block (TB) per transport block set (TBS) on PCH
⇒ 240 bit per transport block

⇒ TTI = 10 ms
⇒ 1/2 rate convolutional coding is used on PCH
⇒ CRC-size = 16 bit

Example: Transport Format Set Definition for PCH-TrCH

The example illustrates the setup of a PCH-TrCH between RNC and NodeB through the NBAP-protocol (NodeB Application Part). The example only shows the very extract of the entire COM_TrCH_SETUP-message that concerns the definition of the PCH.

Example: Transport Format Set Definition for RACH-TrCH

```
+----------+------------------------------------------------+-------------------------------------+
|1.5.1.3.3.1.9 rACH-Parameters                                                                    |
|***B2*** |1.5.1.3.3.1.9.1 id                              |id-RACH-ParametersItem-CTCH-SetupRqs |
|00------ |1.5.1.3.3.1.9.2 criticality                     |reject                               |
|1.5.1.3.3.1.9.3 value                                                                            |
|00110100 |1.5.1.3.3.1.9.3.1 commonTransportChannelID      |52                                   |
|1.5.1.3.3.1.9.3.2 transportFormatSet                                                             |
|1.5.1.3.3.1.9.3.2.1 dynamicParts                                                                 |
|1.5.1.3.3.1.9.3.2.1.1 sequence                                                                   |
|***B2*** |1.5.1.3.3.1.9.3.2.1.1.1 nrOfTransportBlocks     |1                                    |
|***B2*** |1.5.1.3.3.1.9.3.2.1.1.2 transportBlockSize      |168                                  |
|1.5.1.3.3.1.9.3.2.1.1.3 mode                                                                     |
|         |1.5.1.3.3.1.9.3.2.1.1.3.1 notApplicable         |0                                    |
|1.5.1.3.3.1.9.3.2.1.2 sequence                                                                   |
|***B2*** |1.5.1.3.3.1.9.3.2.1.2.1 nrOfTransportBlocks     |1                                    |
|***B2*** |1.5.1.3.3.1.9.3.2.1.2.2 transportBlockSize      |360                                  |
|1.5.1.3.3.1.9.3.2.1.2.3 mode                                                                     |
|         |1.5.1.3.3.1.9.3.2.1.2.3.1 notApplicable         |0                                    |
|1.5.1.3.3.1.9.3.2.2 semi-staticPart                                                              |
|***b3*** |1.5.1.3.3.1.9.3.2.2.1 transmissionTimeInter...  |msec-20                              |
|--01---- |1.5.1.3.3.1.9.3.2.2.2 channelCoding             |convolutional-coding                 |
|-----0-- |1.5.1.3.3.1.9.3.2.2.3 codingRate                |half                                 |
|10010101 |1.5.1.3.3.1.9.3.2.2.4 rateMatcingAttribute      |150                                  |
|-011---- |1.5.1.3.3.1.9.3.2.2.5 cRC-Size                  |v16                                  |
+---------+------------------------------------------------+-------------------------------------+
```

2 Transport Formats:
Transport Format 1:
⇒ 1 transport block (TB) per transport block set (TBS)
⇒ 168 bit per transport block
Transport Format 2:
⇒ 1 transport block (TB) per transport block set (TBS)
⇒ 360 bit per transport block

⇒ TTI = 20 ms
⇒ 1/2 rate convolutional coding is used on RACH
⇒ CRC-size = 16 bit

Example: Transport Format Set Definition for RACH-TrCH

The example illustrates the setup of a RACH-TrCH between RNC and NodeB through the NBAP-protocol (NodeB Application Part). The example only shows the very extract of the entire COM_TrCH_SETUP-message that concerns the definition of the RACH.

Note:
- In the example, two transport formats for the RACH are defined. The respective TFI-values for the two transport formats are not explicitly stated.
- The TFI-values are rather implicitly determined by the position of the transport format in the definition sequence. Therefore, the first transport format with TB-size = 168 bit gets TFI = 0 while the second transport format with TB-size = 360 bit gets TFI = 1.
- This rule applies in general not only between RNC and NodeB (⇔ NBAP-protocol) but also between RNC and UE (⇔ RRC-protocol).
- On PRACH, TFCI = TFI.
- The UE obtains the information about the transport formats on RACH, FACH and other common TrCH's through SIB5.

CCTrCH and Transport Format Combinations (UE Side / FDD)

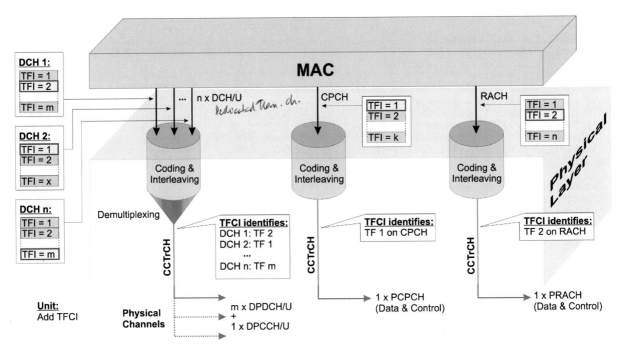

CCTrCH and Transport Format Combinations (UE-Side / FDD)

The CCTrCH
The Coded Composite Transport Channel or CCTrCH was already mentioned earlier. The CCTrCH is applicable only within the physical layer. It represents the unidirectional output data stream of the channel coding and interleaving unit in the physical layer on one hand and the underlying physical channels on the other hand.
⇒ For each type of TrCH (DCH, CPCH and RACH) there is a different CCTrCH.
⇒ Only in case of the DCH-TrCH it is possible that multiple DCH's need to be de-multiplexed on one CCTrCH. This is not possible for the CPCH and the RACH of which there can be only one instance in the UE.

Transport Format Combinations
⇒ As can be seen in the figure, a so called TFCI (Transport Format Combination Identifier) is added to the CCTrCH data stream after the de-multiplexing unit.
⇒ For DCH-TrCH's, the TFCI identifies which particular transport format has been applied during this TTI for each one of the de-multiplexed DCH's within the CCTrCH.
⇒ For CPCH-TrCH's, the TFCI identifies which particular transport format has been applied during this TTI for the CPCH within the CCTrCH.
⇒ For RACH-TrCH's, the TFCI identifies which particular transport format has been applied during this TTI for the RACH within the CCTrCH.

Note:
- If the TTI ranks over multiple radio frames, then the TFCI shall be repeated in all these radio frames.
- Only if multiple DCH-TrCH's are used simultaneously, the TFCI represents the identification of the transport format on each one of the DCH's.
- In case of RACH- and CPCH-TrCH's and if only one DCH-TrCH is configured, the TFCI really only represents the transport format (TFCI = TFI).

⇒ The TFCI will mandatorily be included in the control part of the RACH- and CPCH-physical channel transmission. Note that in case of CPCH and RACH the TFCI-value = TFI-value since there may be only one CPCH or RACH used by the UE.
⇒ The TFCI *may* also be included in the DCH-physical channel transmission, if the appropriate slot format is assigned by the network. Otherwise, the network has to blindly detect the transport formats.

[3GTS 25.302 (6.1)]

Details of Physical Channels (FDD)

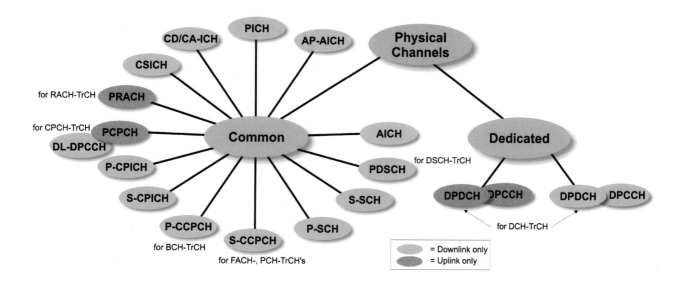

Details of Physical Channels (FDD)

Dedicated Physical Channels
Dedicated physical channels identify the target or destination UE by means of a certain spreading factor and scrambling code.

- **DPDCH**
  Dedicated Physical Data Channel. One or more of these channels can be configured in uplink and downlink direction independently. Each set of DPDCH's is always associated with the

- **DPCCH** (Dedicated Physical Control Channel)

Common Physical Channels
On common physical channels that carry common TrCH's, the target or source UE needs to be identified by means of an inband identifier (⇔ RNTI).

- **AP-AICH**
  CPCH Access Preamble Acquisition Indicator Channel
- **CPCH**
  Common Packet Channel
- **PICH**
  Paging Indicator Channel
- **CD/CA-ICH**
  Collision Detection / Channel Assignment Indicator Channel
- **CSICH**
  CPCH Status Indicator Channel
- **PRACH**
  Physical Random Access Channel
- **PCPCH**
  Physical Common Packet Channel
- **P-CPICH**
  Primary Common Pilot Channel
- **S-CPICH**
  Secondary Common Pilot Channel
- **P-CCPCH**
  Primary Common Control Physical Channel
- **S-CCPCH**
  Secondary Common Control Physical Channel
- **P-SCH**
  Primary Synchronization Channel
- **S-SCH**
  Secondary Synchronization Channel
- **PDSCH**
  Physical Downlink Shared Channel
- **AICH**
  Acquisition Indicator Channel

[3GTS 25.211 (5)]

Layer 1 Signaling in Data Transferring Physical Channels

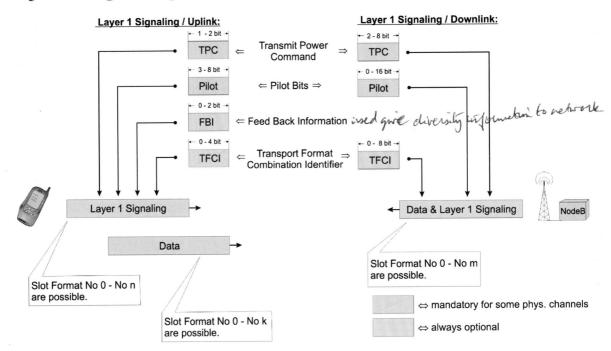

Layer 1 Signaling in Data Transferring Physical Channels

On all physical channels that are available for data exchange between the UE and the NodeB there exists the need to exchange layer 1 signaling information between UE and NodeB. Which type of signaling information shall be used and how it is to be interpreted by UE and NodeB is selected by the RRC-layer during the setup of the respective physical channels. Accordingly, these physical channels support different slot formats which are identified through the presence and length of the layer 1 signaling information and through the spreading factor.

The following layer 1 information is defined:

TPC-Field
The TPC-field (Transmit Power Command) is mandatory in uplink and downlink direction. However, the length of the TPC-field is variable and depends on the selected slot format. The TPC-field is used for fast closed loop power control and tells the receiver to either increase or decrease its transmit power.

Pilot Bits
Pilot bits consist of a pre-defined bit sequence to allow the receiver a channel quality estimation. The number of pilot bits per slot is variable in uplink and downlink direction and depends on the slot format as selected by RRC. On S-CCPCH, the presence of pilot bits is optional.

FBI-Field
The FBI-field (FeedBack Information) is optionally used in uplink direction only. The FBI-field is again split into the S-field and the D-field:
⇒ The S-field is used to select the primary cell in a soft-handover situation, if the *site selection diversity transmit power control* procedure (SSDT) is used. All non-primary cells shall switch-off their downlink transmission to that mobile station.
⇒ The D-field is used to provide phase shift feedback information to the NodeB for both transmit antennas in case of *closed loop mode transmit diversity*.

TFCI-Field
The TFCI-field is used to inform the receiver of a physical channel about the transport formats of the embedded transport channels. In the control part of the PRACH and PCPCH, the presence of the TFCI-field is mandatory / fixed length while on DPDCH, the presence and length of the TFCI-field depends on the slot format as selected by RRC.

[3GTS 25.211 (5.2.1), 3GTS 25.214 (5.2.1.4 / 7)]

Scrambling code is used to identify the transmitter

Example: Slot Format Definition for DPDCH/U and DPCCH/U

4. Implicit selection of non-compressed format because "compressed format IE" is not present in the RB_SETUP- message. Therefore transmission will occur on 15 slots per radio frame

1. SF = 64 only relates to the minimum spreading factor and therefore the slot format on DPDCH/U. For DPCCH/U there will always be SF = 256

Extract of 3GTS 25.211 (Table 11 / Slot Format DPCCH/U):

Slot Format #i	Channel Bit Rate (kbps)	Channel Symbol Rate ksps	SF	Bits/Frame	Bits/Slot	N_{pilot}	N_{TPC}	N_{TFCI}	N_{FBI}	Transmitted slots per radio frame
0	15	15	256	150	10	6	2	2	0	15
0A	15	15	256	150	10	5	2	3	0	10-14
0B	15	15	256	150	10	4	2	4	0	8-9
1	15	15	256	150	10	8	2	0	0	8-15
2	15	15	256	150	10	5	2	2	1	15
2A	15	15	256	150	10	4	2	3	1	10-14
2B	15	15	256	150	10	3	2	4	1	8-9
3	15	15	256	150	10	7	2	0	1	8-15
4	15	15	256	150	10	6	2	0	2	8-15
5	15	15	256	150	10	5	1	2	2	15
5A	15	15	256	150	10	4	1	3	2	10-14
5B	15	15	256	150	10	3	1	4	2	8-9

Extract of RB_SETUP-Message

```
                   2.1.1.1.9    ul-ChannelRequirement
                   2.1.1.1.9.1  ul-DPCH-Info
                   2.1.1.1.9.1.1   modeSpecificInfo
                   2.1.1.1.9.1.1.1 fdd
---1----           2.1.1.1.9.1.1.1.1 scramblingCodeType     longSC
**b24***           2.1.1.1.9.1.1.1.2 scramblingCode         1084104
                   2.1.1.1.9.1.1.1.3 numberOfDPDCH          1
----100-           2.1.1.1.9.1.1.1.4 spreadingFactor        sf64
-------1           2.1.1.1.9.1.1.1.5 tfci-Existence         1
```

3. TFCI shall be present.

10 ms ~ 38400 chips

2. The IE "Number of FBI-bits" is not present after the IE "TFCI-Existence". Accordingly, the related slot format shall contain no FBI-bits.

— We have 12 different slot formats which MS has to handle.

Example: Slot Format Definition for DPDCH/U and DPCCH/U

The graphics pages illustrates an extract of an RB_SETUP-message. This message contains, among other things, the IE "Uplink DPCH-Info" which informs the UE which slot format it shall apply on the respective DPDCH/U and DPCCH/U.
The green field illustrates an extract of 3GTS 25.211 that defines the different slot formats for DPCCH/U.

Note: Unfortunately, the slot format to be used is not included in the RRC: RB_SETUP-message in plain text. It needs to be determined by an iterative procedure that is explained underneath. This procedure applies only between RNC and UE, since the RNC *will* convey the slot format to be used through the NBAP-protocol to the NodeB in plain text.

The following numbers relate to the numbers in the text boxes on the previous page.

1. The IE "spreadingFactor" only relates to the minimum spreading factor that the mobile station may use on DPDCH/U. The spreading factor of the DPCCH/U is always SF = 256. Consequentially, on DPDCH/U the mobile station may only use slot format 0 ($\Leftrightarrow$ SF = 256 / 15 kbit/s), slot format 1 ($\Leftrightarrow$ SF = 128 / 30 kbit/s) and slot format 2 ($\Leftrightarrow$ SF = 64 / 60 kbit/s).
2. The IE "Number of FBI-bits" is missing. Implicitly, this means that the mobile station shall not include any FBI-bits (Feed Back Information). Consequentially, on DPCCH/U only slot formats without FBI-bits are applicable.
3. The network instructs the mobile station to include a TFCI-field into the DPCCH/U.
4. No compressed mode operation is applicable. Accordingly, the mobile station shall always transmit all 15 slots per radio frame.

Result: The only slot format that fulfills all previous constraints, is slot format #0.

[3GTS 25.211 (table 1, table 2), 3GTS 25.331 (10.3.6.88)]

Example: Slot Format Definition for DPDCH/D and DPCCH/D

Extract of RRC_CONN_SETUP-Message

Result: Consequentially, slot format # 4 is selected on DPDCH/D + DPCCH/D.

```
|1.1.1.1.13     dl-CommonInformation
|1.1.1.1.13.1   dl-DPCH-InfoCommon
|1.1.1.1.13.1.1 cfnHandling
|1.1.1.1.13.1.1.1 initialise
|1.1.1.1.13.1.2 modeSpecificInfo
|1.1.1.1.13.1.2.1 fdd
|1.1.1.1.13.1.2.1.1 dl-DPCH-PowerControlInf
|1.1.1.1.13.1.2.1.1.1 modeSpecificInfo
|1.1.1.1.13.1.2.1.1.1.1 fdd
|-----0--  |1.1.1.1.13.1.2.1.1.1.1.1 dpc-Mode          |singleTPC
|***b5***  |1.1.1.1.13.1.2.1.2 powerOffsetPilot-pdpdch |0
|         |1.1.1.1.13.1.2.1.3 spreadingFactorAndPilot
|------01 |1.1.1.1.13.1.2.1.3.1 sfd256                 |pb4
|0------- |1.1.1.1.13.1.2.1.4 positionFixedOrFlexible  |fixed
|-0------ |1.1.1.1.13.1.2.1.5 tfci-Existence           |0
```

1. Implicit selection of non-compressed format because "compressed format IE" is not present in the above message. Therefore transmission will occur on 15 slots per radio frame

2. SF = 256 (⇔ applies to both DPDCH/D and DPCCH/D).

3. TFCI will not be present on DPCCH. (Default would be present)

4. The number of pilot bits shall be 4 per slot (⇔ pb4)

Extract of 3GTS 25.211 (Table 11):

Slot Format #i	Channel Bit Rate (kbps)	Channel Symbol Rate	SF	Bits/Slot	DPDCH Bits/Slot		DPCCH Bits/Slot			Transmitted slots per
					N_{Data1}	N_{Data2}	N_{TPC}	N_{TFCI}	N_{Pilot}	
0	15	7.5	512	10	0	4	2	0	4	15
0A	15	7.5	512	10	0	4	2	0	4	8-14
0B	30	15	256	20	0	4	0	8	8-14	
1	15	7.5	512	10	0	2	2	2	4	15
1B	30	15	256	20	0	4	4	4	8	8-14
2	30	15	256	20	2	14	2	0	2	15
2A	30	15	256	20	2	14	2	0	2	8-14
2B	60	30	128	40	4	28	4	0	4	8-14
3	30	15	256	20	2	12	2	2	2	15
3A	30	15	256	20	2	10	2	4	2	8-14
3B	60	30	128	40	4	24	4	4	4	8-14
4	30	15	256	20	2	12	2	0	4	15
4A	30	15	256	20	2	12	2	0	4	8-14
4B	60	30	128	40	4	24	4	0	8	8-14
5	30	15	256	20	2	10	2	2	4	15
5A	30	15	256	20	2	8	2	4	4	8-14
5B	60	30	128	40	4	20	4	4	8	8-14
6	30	15	256	20	2	8	2	0	8	15

(table continues)

Example: Slot Format Definition for DPDCH/D and DPCCH/D

The graphics pages illustrates an extract of an RRC_CONN_SETUP-message. This message contains, among other things, common information for all configured downlink channels (⇔ dl-Common-Information) which in turn contains information about the very slot format on DPDCH/D and DPCCH/D that will be applied by the NodeB.
The green field illustrates an extract of 3GTS 25.211 that defines the different slot formats for DPDCH/D and DPCCH/D.

> Note: Unfortunately, the slot format to be used is not included in the RRC: RRC_CONN_REQ-message in plain text. It needs to be determined by an iterative procedure that is explained underneath. This procedure applies only between RNC and UE, since the RNC *will* convey the slot format to be used through the NBAP-protocol to the NodeB in plain text.

The following numbers relate to the numbers in the text boxes on the previous page.

5. The IE "modeSpecificInfo" does not include the optional IE "compressedInfo". Accordingly, compressed mode operation is not applicable. Consequentially, all slot formats where less than 15 slots per frame shall be transmitted, are not applicable.
6. The IE "Spreading Factor and Pilot" clearly identifies that a spreading factor 256 will be used. Since DPDCH/D and DPCCH/D are time-multiplexed, this spreading factor applies to both channels. Consequentially, only slot formats which use SF = 256 are applicable.
7. The network will not include a TFCI-field. The non-existence of the TFCI-field does require blind detection to identify the different transport formats of the embedded DCH. Either way, the potentially possible slot formats that include a TFCI-field are excluded.
8. The final identification is given by the information that 4 pilot bits shall be included per slot.

Result: The only slot format that fulfills all previous constraints, is slot format #4.

[3GTS 25.211 (table 11), 3GTS 25.331 (10.3.6.24, 10.3.6.18)]

Logical Format of Physical Channels (FDD Uplink)

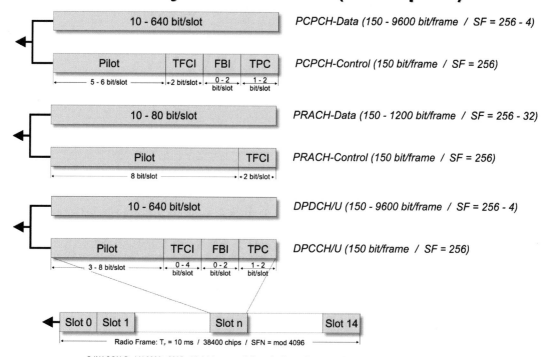

Logical Format of Physical Channels (FDD Uplink)

PCPCH-Data and Control (Physical Common Packet Channel)
The PCPCH is exclusively used as bearer for the CPCH-TrCH. The data part carries 150 – 9600 channel coded and interleaved data bits per radio frame (10 ms) from the UE to NodeB. The throughput rate is therefore 15 kbit/s – 960 kbit/s. The spreading factor is either 256, 128, 64, 32, 16, 8 or 4 ($\Leftrightarrow$ SF = $256 / 2^k$ with k = 0 ... 6). There are 7 different slot formats defined for the PCPCH-data part and identified by the spreading factor. For the PCPCH-control part there are 3 different slot formats defined, which differ in the number of bits for the Pilot-, FBI-, and TPC-fields. The slot formats for the data and control part are pre-configured by the RNC and provided to the UE in system information messages. The very transport format of the CPCH which is used by the mobile station during one radio frame is signaled through the TFCI-field (30 bit / radio frame).

PRACH-Data and Control (Physical Random Access Channel)
The PRACH is exclusively used as bearer for the RACH-TrCH. The data part carries 150 – 1200 channel coded and interleaved data bits per radio frame (10 ms) from the UE to NodeB. Only TTI-values 10 ms and 20 ms are applicable on the RACH. Applicable transport formats and TTI-value are broadcast in SIB 5. The theoretical throughput rate of the PRACH is 15 kbit/s – 120 kbit/s, however any given UE may use the RACH-TrCH (and therefore the PRACH) only during one TTI. For another data transmission a new random access procedure is required. Note that on RACH-TrCH only the open loop power control procedure is applicable. The spreading factor of the PRACH is either 256, 128, 64 or 32 ($\Leftrightarrow$ SF = $256 / 2^k$ with k = 0 ... 3). For the PRACH-data part, 4 different slot formats are defined and identified through the spreading factor of the PRACH in SIB5. Only one slot format is defined for the control part. The very transport format of the RACH which is used by the mobile station during one radio frame is signaled through the TFCI-field (30 bit / radio frame) of the PRACH.

DPDCH/U (Dedicated Physical Data Channel / Uplink)
One or more DPDCH/U are exclusively used as bearer for one or more DCH/U-TrCH's. The data part carries 150 – 9600 channel coded and interleaved data bits per radio frame (10 ms) from the UE to NodeB. The throughput rate is therefore 15 kbit/s – 960 kbit/s. The spreading factor is either 256, 128, 64, 32, 16, 8 or 4 ($\Leftrightarrow$ SF = $256 / 2^k$ with k = 0 ... 6). For the DPDCH/U-data part, 7 different slot formats are defined and identified through the spreading factor parameter which is part of the IE "UL-DPCH-Info" in the respective allocation message.

DPCCH/U (Dedicated Physical Control Channel / Uplink)
The DPCCH/U carries the layer 1 signaling information for one or more DPDCH/U. The DPCCH/U carries 150 bits per radio frame (10 ms) from the UE to the NodeB. The throughput rate is therefore 15 kbit/s, the spreading factor is SF = 256. For the DPCCH/U, 12 different slot formats are defined. They differ in the presence and length of the different embedded parameters (Pilot, TFCI, FBI and TPC). The selection of the very slot format to be used is done by the SRNC ($\Leftrightarrow$ RRC-protocol) and signaled to the UE in the respective allocation message (IE "UL-DPCH-Info / see previous example).

[3GTS 25.211 (5.2)]

(1) Logical Format of Physical Channels (FDD Downlink)

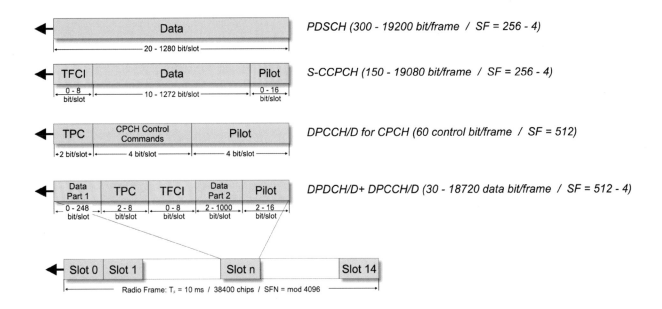

(1) Logical Format of Physical Channels (FDD Downlink)

PDSCH (Physical Downlink Shared Channel)
The PDSCH is exclusively used as bearer for the DSCH-TrCH. The PDSCH carries 300 – 19200 channel coded and interleaved data bits per radio frame (10 ms) from the NodeB to the UE. The throughput rate is therefore 30 kbit/s – 1920 kbit/s. The spreading factor is either 256, 128, 64, 32, 16, 8 or 4 ($\Leftrightarrow$ SF = 256 / 2^k with k = 0 ... 6). Note that there is no layer 1 signaling information included in the PDSCH. The primary or one related secondary scrambling code is used for the PDSCH.

S-CCPCH (Secondary Common Control Physical Channel)
The S-CCPCH is used as bearer for the FACH- and PCH-TrCH's. The number of channel coded and interleaved data bits per radio frame varies between 150 bit minimum and 19080 bit maximum, depending on the slot format selected by the CRNC. Accordingly, the data throughput rate varies between 15 kbit/s and 1908 kbit/s. The slot format also determines whether a TFCI-field and / or pilot bits shall be included in the S-CCPCH. Altogether, 18 different slot formats have been defined. The very slot format which is used for S-CCPCH is broadcast by the NodeB in SIB5 by identifying how many bits shall be used for TFCI and pilot and it is identified through the spreading factor. The primary scrambling code of the cell is used for the S-CCPCH, if it carries the PCH. If it only carries FACH, one related secondary scrambling code can be used.

DPCCH/D for CPCH (Dedicated Physical Control Channel)
The DPCCH/D for CPCH is exclusively used to carry higher layer control information and layer 1 signaling data from the NodeB to the UE for the uplink only CPCH-TrCH. There is only one slot format defined with a fixed spreading factor of 512. When we consider the CPCH control commands as signaling payload of the DPCCH/D for CPCH, the throughput rate is 6 kbit/s.

DPDCH/D and DPCCH/D (Downlink DPCH)
One or more DPDCH/D together with a single DPCCH/D are exclusively used as bearer for one or more DCH/D-TrCH's. The data part carries 30 – 18720 channel coded and interleaved data bits per radio frame (10 ms) from the NodeB to the UE. The throughput rate is therefore 3 kbit/s – 1872 kbit/s. The spreading factor is either 512, 256, 128, 64, 32, 16, 8 or 4 ($\Leftrightarrow$ SF = 512 / 2^k with k = 0 ... 7). The presence and length of the various fields of the DPDCH/D+DPCCH/D depends on the very slot format to be used. Altogether 49 slot formats have been defined. The slot format to be used by the NodeB is selected by the SRNC ($\Leftrightarrow$ RRC-protocol) and signaled to Node B and UE in the respective allocation message (IE "DL-Common-Info / see previous example). The primary or one related secondary scrambling code is used for the DPCCH/D and the DPDCH/D. In case of compressed mode transmission, alternative scrambling codes are used.

[3GTS 25.211 (5.3)]

(2) Logical Format of Physical Channels (FDD Downlink)

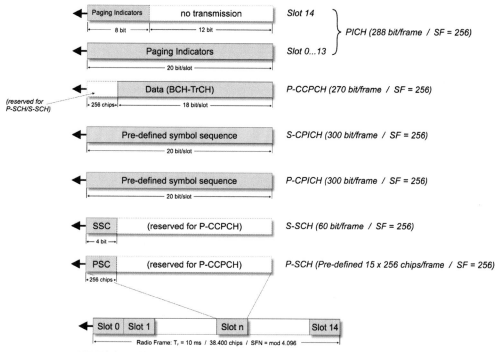

(2) Logical Format of Physical Channels (FDD Downlink)

PICH (Paging Indicator Channel)
The PICH is used to transmit between 18 and 144 paging indicators per radio frame to all surrounding UE's. The paging indicators will tell the mobile station whether they shall listen to an S-CCPCH to receive a paging. The PICH is always transmitted using the primary scrambling code of the cell (⇔ primary SC = No 0, No 16, No 32, ... No 8176).

P-CCPCH (Primary Common Control Physical Channel)
The P-CCPCH is exclusively used as bearer for the BCH-TrCH. The P-CCPCH carries 270 channel coded and interleaved data bits per radio frame (10 ms) from the NodeB to the UE. The throughput rate is therefore 27 kbit/s. Note that on all UTRA-cells the P-CCPCH uses the same spreading factor of length 256 chips (SF = 256). The P-CCPCH is switched off during the first 256 chips (⇔ 10 %). This time period is used by the P-SCH and S-SCH which are therefore time-synchronized with the P-CCPCH. The P-CCPCH is always transmitted using the primary scrambling code of the cell (see PICH).

S-CPICH (Secondary Common Pilot Channel)
The S-CPICH is an optional physical channel that may be used as a phase reference for an S-CCPCH. In that respect, the S-CPICH will use a spreading factor which is allocated by the network and either the primary or a related secondary scrambling code.

P-CPICH (Primary Common Pilot Channel)
The P-CPICH broadcasts a pre-defined bit sequence to the surrounding UE's. The P-CPICH always uses the same spreading factor of length 256 chips on all UTRA-cells. Note however, that this spreading factor is different to the one which is used on P-CCPCH. Most importantly, the P-CPICH is the phase reference for many other physical channels plus it serves as beacon for neighbor cell and serving cell measurements. The P-CPICH is always transmitted using the primary scrambling code of the cell (see PICH).

S-SCH (Secondary Synchronization Channel)
The S-SCH is used to broadcast the very scrambling code group of a cell towards all surrounding UE's. Therefore, the 512 different primary downlink scrambling codes (⇔ primary SC = No 0, No 16, No 32, ... No 8176) are divided into 64 code groups of which each group contains 8 different scrambling codes. Each code group is identified on S-SCH through a unique symbol sequence over one radio frame. After the correct scrambling code group has been detected, the UE has to probe each of the 8 included scrambling codes to be able to decode the P-CCPCH.

P-SCH (Primary Synchronization Channel)
The P-SCH is used to transmit the 256 chips long PSC (Primary Synchronization Code) in every slot of every radio frame. The PSC is identical for each UTRA-cell and is used for time synchronization with the simultaneously transmitted S-SCH and the consecutively transmitted P-CCPCH. The chips of the PSC will be inverted before modulation, if STTD-encoding is used on P-CCPCH.

[3GTS 25.211 (5.3)]

(3) Logical Format of Physical Channels (FDD Downlink)

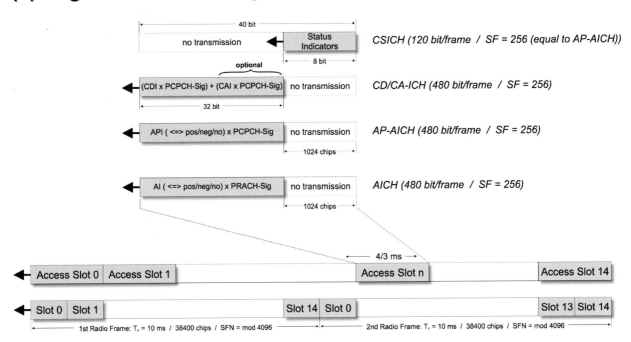

(3) Logical Format of Physical Channels (FDD Downlink)

Note that the following channels use access slots for transmission which have twice the duration of a usual slot.

CSICH (CPCH Status Indicator Channel)
The CSICH is used to inform the listening UE's about the availability and optionally the available spreading factors of the different CPCH's. If a CPCH is already in use, the respective status indicator will tell a UE the "busy"-condition. Therefore, the UE obtains information beforehand whether a CPCH-access procedure may be successful or not. The CSICH is always related with the AP-AICH and it is using the same spreading code (SF = 256). The throughput rate of the CSICH is 6 kbit/s. The CSICH is always transmitted using the primary scrambling code of the cell (see PICH).)

CD/CA-ICH (CPCH Collision Detection / Channel Assignment Indicator Channel)
The CD/CA-ICH is used during the CPCH-access procedure to verify that only one UE will consecutively transmit on PCPCH. The CD/CA-ICH shall approve the PCPCH-signature of the UE and may optionally include a CPCH channel assignment indicator. The CD/CA-ICH is always transmitted using the primary scrambling code of the cell (see PICH). The spreading factor is SF = 256, the throughput rate is 24 kbit/s.

AP-AICH (CPCH Access Preamble Acquisition Indicator Channel)
The APAICH is used to carry access preamble acquisition indicators (API) for CPCH-TrCH to all listening UE's during the CPCH-access procedure. There are 16 different API's of length 32 bit and each API corresponds to exactly one PCPCH-signature. An API can either be positive (API x (+1)), negative (API x (−1) / inverted AI) or non existent ('0'). The AP-AICH is always transmitted using the primary scrambling code of the cell (see PICH). The time with no transmission in each slot may be used by the CSICH. The spreading factor is SF = 256, the throughput rate is 24 kbit/s.

AICH (Acquisition Indicator Channel)
The AICH is used to carry acquisition indicators (AI) to all listening UE's during the RACH-access procedure. There are 16 different AI's of length 32 bit and each AI corresponds to exactly one PRACH-signature. An AI can either be positive (AI x (+1)), negative (AI x (−1) / inverted AI) or non existent ('0'). The AICH is always transmitted using the primary scrambling code of the cell (see PICH). The time with no transmission in each slot may be used by the CSICH. The spreading factor is SF = 256, the throughput rate is 24 kbit/s.

[3GTS 25.211 (5.3)]

UMTS – Signaling & Protocol Analysis

- *A Comprehensive Inside View on UMTS*
- **Signaling & Protocol Analysis in UTRAN**
- *Important UMTS Scenarios & Call Tracing*

UMTS – Signaling & Protocol Analysis

Intentionally left blank

Objectives

After this Lecture the Student will:

- **Understand the various MAC-Tasks and Procedures in Detail**

- **Know how RLC Offers Different Operation Modes (TM, UM and AM) to Higher Layers and the Related Procedures**

- **Understand the RRC-Protocol and the Related Functions in Detail**

- **Obtain an Overview of ASN.1 Basic and Packed Encoding Rules**

Objectives

- **Understand the various MAC-Tasks and Procedures in Detail**
  Among other things, MAC takes care of the mapping between logical channels and transport channels. We will elaborate in detail which mappings are possible.

- **Know how RLC Offers Different Operation Modes (TM, UM and AM) to Higher Layers and the Related Procedures**
  The layer on top of MAC is RLC, a classical link layer implementation that offers different operation modes to match the highly variable QoS-requirements within UMTS. We will focus on the different RLC-PDU-types and highlight the peer-to-peer procedures within RLC.

- **Understand the RRC-Protocol and the Related Functions in Detail**
  The most important protocol within UTRAN is the RRC-protocol. Most other protocols are directly or indirectly controlled by the RRC-protocol. As the name "radio resource control" suggests, RRC takes care of all issues and tasks that relate to the UE accessing the UTRAN radio resources.

- **Obtain an Overview of ASN.1 Basic and Packed Encoding Rules**
  Although RRC is not the only protocol in UMTS that uses the Abstract Syntax Notation 1 for message and information element encoding, we like to present this complex description language before our consideration of the various RRC-messages. However, the basic output product of ASN.1 encoding is not very efficient and produces a lot of redundant information. Therefore, the UTRAN protocols RRC, NBAP, RANAP and RNSAP deploy in addition the PER (Packed Encoding Rules) that streamline the message code which is produces by ASN.1

Medium Access Control (MAC)

- Tasks and Functions

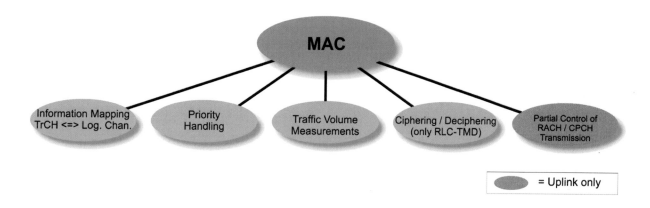

Medium Access Control (MAC)

Overview: Tasks and Functions

- **Information Mapping TrCH ⇔ Logical Channel**
  As per configuration from RRC, the MAC-layer handles the mapping between logical channels and transport channels. This mapping includes the selection of a suitable transport format depending on the size and number of the received RLC-PDU's. We will illustrate the mapping of logical channels to transport channels and of transport channels to physical channels in full detail for FDD-mode from the UE's perspective.

- **Priority Handling**
  MAC is responsible to multiplex different logical channels on the same TrCH, if required. MAC is also responsible to prioritize the transmission of PDU's from these logical channels over other logical channels based on the MLP (MAC Logical Channel Priority). In the network side and for common transport channels, this prioritization also applies for multiplexing MAC-PDU's destined for different UE's on a single TrCH.

- **Traffic Volume Measurement**
  MAC is responsible to provide traffic volume measurement reports to the RRC-layer to provide for the adjustment of the related TrCH-capacities. In that respect, MAC will receive from RLC the current BO (Buffer Occupancy) and related parameters for each logical channel that uses a configured TrCH. This information is provided to MAC at least once every 10 ms.

- **Ciphering / Deciphering (only RLC-TMD)**
  RLC takes care of ciphering and deciphering RLC-UMD- and RLC-AMD-PDU's. However, the encryption of RLC-TMD-PDU's is a MAC-function.

  > Note: Ciphering for RLC-TMD-PDU's is not applicable on CCCH (⇔ Signaling Radio Bearer 0).

- **Partial Control of RACH / CPCH Transmissions**
  MAC receives control information from RRC about the priority of the data to be sent on RACH or CPCH. This priority information is used by MAC to control most of the random access procedures for RACH- and CPCH-transmission. However, the timing on slot level is controlled by the physical layer.

[3GTS 25.321 (6.1)]

Mapping of Logical, Transport and Physical Channels (FDD)

- Mapping of BCCH and PCCH (UE-Perspective)

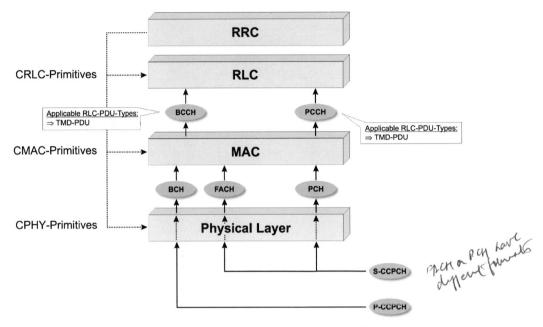

Mapping of Logical, Transport and Physical Channels (FDD)

The mapping of logical channels to transport channels and of transport channels to physical channels is not random, however, there are many options particularly for the mapping between logical channels and transport channels. Which mapping is actually used is selected and configured by the RRC-protocol through primitives between RRC and the respective layer. There are CRLC-Primitives between RRC and RLC, CMAC-Primitives between RRC and MAC and CPHY-Primitives between RRC and the physical layer. This page and the following pages contain a complete presentation of the possible mapping for all channels which are defined for UTRA-FDD-mode from the UE's perspective.

Mapping of BCCH and PCCH (UE-Perspective)
⇒ The BCCH can only be mapped on either the BCH-TrCH or on one FACH-TrCH. One layer below, the BCH-TrCH can only be mapped on P-CCPCH while the FACH-TrCH can only be mapped on S-CCPCH.
⇒ Only RLC-TMD-PDU's can be received on BCCH.

The only RRC-messages to be sent over the BCCH are the SYS_INFO- and SYS_INFO_CHANGE_IND-messages.

⇒ The PCCH can only be mapped on the PCH-TrCH. Like the FACH-TrCH, the PCH-TrCH will be mapped on the same or another S-CCPCH than the FACH-TrCH.
⇒ Only RLC-TMD-PDU's can be received on PCCH.

The only RRC-message to be sent over the PCCH is the PAG_TYPE1-message.

[3GTS 25.211 (6.1), 3GTS 25.301 (5.3.1.1.2)]

Mapping of DCCH and DTCH (UE-Perspective)

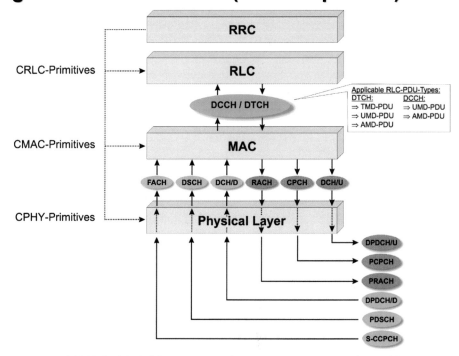

Mapping of DCCH and DTCH (UE-Perspective)

- ⇒ In downlink direction, the DCCH and DTCH can be mapped on common transport channels (FACH-, DSCH- TrCH) and on dedicated transport channels (DCH-TrCH). In uplink direction, the DCCH and DTCH can be mapped on common transport channels (RACH-, CPCH-TrCH) and on dedicated transport channels (DCH-TrCH).
- ⇒ Which mapping is actually used, depends on the configuration primitives from RRC.
- ⇒ On the DCCH, only UMD- and AMD-RLC-PDU's can be sent and received while on DTCH also RLC-TMD-PDU's are applicable.

On DCCH, only those RRC-messages can be sent which are configured for the DCCH in uplink and downlink direction.

[3GTS 25.211 (6.1), 3GTS 25.301 (5.3.1.1.2)]

Mapping of CCCH (UE-Perspective)

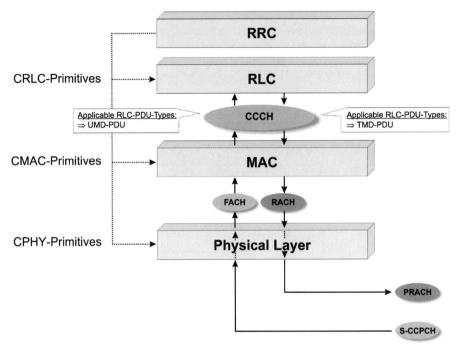

Mapping of CCCH (UE-Perspective)

⇒ In downlink direction, the CCCH can only be mapped on FACH-TrCH's (common transport channel). Only RLC-UMD-PDU's can be sent on the CCCH in downlink direction.

On CCCH in downlink direction, only 5 different RRC-messages can be sent (CELL_UPD_CNF, RRC_CONN_REJ, RRC_CONN_REL, RRC_CONN_SETUP, URA_UPD_CNF).

⇒ In uplink direction, the CCCH can only be mapped on the RACH-TrCH (common transport channel). Only RLC-TMD-PDU's can be sent on the CCCH in uplink direction.

On CCCH in uplink direction, only 3 different RRC-messages can be sent (CELL_UPD, RRC_CONN_REQ, URA_UPD).

[3GTS 25.211 (6.1), 3GTS 25.301 (5.3.1.1.2)]

Mapping of CTCH (UE-Perspective)

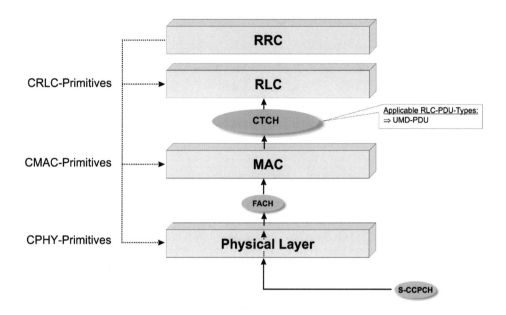

Mapping of CTCH (UE-Perspective)

⇒ The CTCH is a PTM-channel (Point-to-Multipoint) that only exists in downlink direction and only in the user plane. It can only be mapped on the FACH-TrCH. Only RLC-UMD-PDU's can be sent on the CTCH.
⇒ The CTCH is mostly used to convey BMC-PDU's to the UE's.

[3GTS 25.211 (6.1), 3GTS 25.301 (5.3.1.1.2)]

MAC-PDU Formats on Logical Channels (FDD / UE-Side)

- Format on BCCH, PCCH and CTCH

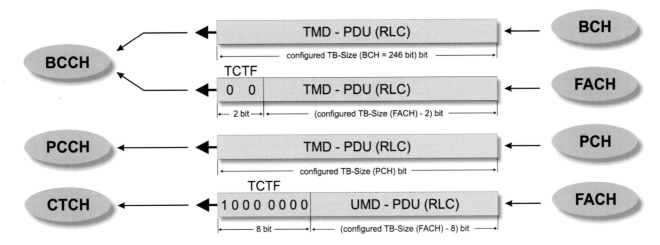

MAC-PDU Formats on Logical Channels (FDD / UE-Side)

Depending on the combination logical channel ⇔ transport channel, MAC-PDU's need to include more or less control information (⇔ MAC-header) to provide for a proper routing of that MAC-PDU at the receiver side. This may include the provision of so called RNTI's (Radio Network Temporary Identifiers) to identify the affected UE. It is also essential to know which logical channel ⇔ transport channel combination allows for which RLC-PDU-type (TMD, UMD, AMD) to be embedded. We illustrate these combinations from the perspective of the UE. Obviously, the perspective of the UTRAN is vice versa.

Note:
- The size of a MAC-PDU depends on the previously configured transport block size (TB-size) for the used TrCH, as the figures on this page and the following pages illustrate.
- Each MAC-PDU is represented by exactly one transport block.

Format on BCCH, PCCH and CTCH

- **BCCH**
  On BCCH, only RLC-TMD-PDU's are allowed. If the BCCH is mapped on BCH-TrCH, then there is no MAC-header added to the TMD-PDU. Note that the transport block size on BCH is always 246 bit and accordingly, MAC-PDU's on BCH will have a size of 246 bit.
  The FACH as TrCH for the BCCH is only applicable for SIB 10 and shall only be read by UE's which support the simultaneous reception of one or more DPDCH/D and the S-CCPCH. In this case, the MAC-header consists of the TCTF-field (Target Channel Type Field). On FACH, the TCTF-field can have a variable length of 2 bit or 8 bit. The bit combination '00'(bin) identifies the BCCH as destined logical channel.

- **PCCH**
  The PPCH can only be mapped to the PCH-TrCH. Only RLC-TMD-PDU's are allowed. No MAC-header added to these TMD-PDU's. The size of each MAC-PDU on PCH depends on the previously configured TB-sizes for PCH that the mobile station determines out of system information messages (⇔ SIB 5). The same applies for the FACH-TrCH.

- **CTCH**
  The only allowed TrCH for the CTCH is the FACH. To provide for a proper mapping routing of received FACH-TrCH blocks to CTCH, the TCTF-field is needed. MAC-PDU's which are destined for CTCH are identified by the 8 bit bit-pattern '1000 000'(bin). Only RLC-UMD-PDU's can be embedded into the respective MAC-PDU's.

[3GTS 25.321 (9.2.1)]

Format on CCCH

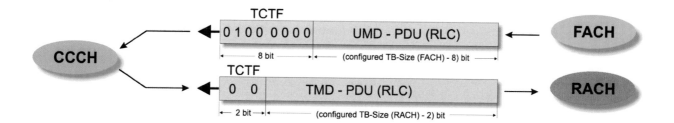

Format on CCCH

⇒ The CCCH logical channel is bi-directional logical channel.
⇒ In uplink direction, the only applicable TrCH is the RACH.
⇒ In downlink direction, the only applicable TrCH is the FACH.
⇒ Since RACH and FACH can also be used as bearers for other logical channels, a TCTF-field is required to identify the CCCH as destination.
⇒ On RACH, the TCTF-field has always a size of 2 bit (mandatorily present, irrespective of the logical channel). The bit combination TCTF = '00'(bin) uniquely identifies the CCCH as destination. Note that the size of the MAC-PDU's on RACH is given the pre-configured TB-sizes for the RACH-TrCH. The MAC-PDU's on RACH for CCCH may only contain RLC-TMD-PDU's.
⇒ On FACH for CCCH, an 8 bit long TCTF-field is used to identify the CCCH as destination (TCTF = '0100 0000'(bin)). Note that the size of the MAC-PDU's on FACH is given the pre-configured TB-sizes for the FACH-TrCH. The MAC-PDU's on FACH for CCCH may only contain RLC-UMD-PDU's.

Note: The TCTF-field in MAC-PDU's on FACH is frequently 8 bit long although 2 bit would be sufficient. This is always true, if the FACH is exclusively used to transfer RLC-UMD-PDU's because the size of UMD-PDU's (and AMD-PDU's) needs to be a multiple of 8 bit. Considering that the RLC-PDU-size equals the TB-size – MAC-header size, the MAC-header needs to have a length of a multiple of 8 bit in these cases, provided that the configured TB-size has a length of a multiple of 8 bit.

[3GTS 25.321 (9.2.1)]

Format on DTCH and DCCH

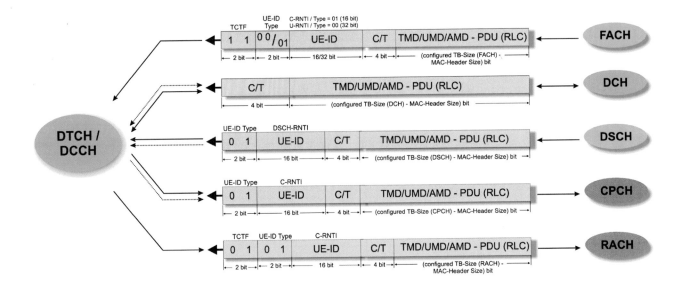

Format on DTCH and DCCH

Note:
- The most complex mapping between logical channels and transport channels applies, if dedicated control or traffic channels (DCCH / DTCH) are mapped on common or shared transport channels in uplink or downlink direction.
- This is obvious: Common TrCH's are not related to a single UE. Accordingly, the target or source UE needs to be identified in the MAC-header to avoid ambiguities. This is achieved through the inclusion of different types of RNTI's.
- If more than one logical channel is mapped on a TrCH, it is also necessary to identify the logical channel number. This requirement is taken care of by the 4 bit long C/T-field (⇔ up to 16 logical channels can be distinguished per transport channel).
- MAC-PDU's for DCCH and DTCH may contain all three types of RLC-PDU's (TMD, UMD and AMD). The very type to be used depends on the QoS and is configured by RRC.

⇒ When DTCH or DCCH are mapped on a common FACH-TrCH, the MAC-header consists of the TCTF-field, an identification which type of RNTI (⇔ UE-ID) is used to identify the UE, the RNTI itself and the C/T-field. Note that the C/T-field is mandatory, because there will always be multiple DCCH-/DTCH-logical channels defined that need to be distinguished per UE. Whether C-RNTI or U-RNTI are used as UE-ID, depends on whether a C-RNTI has already been allocated and conveyed to the UE. Initially, the CRNC needs to use the U-RNTI (possibly allocated by a different SRNC) to address the UE.
⇒ When DTCH or DCCH are mapped on a dedicated DCH-TrCH, no identification of the UE is required and the MAC-header consists only of the optionally present C/T-field. The C/T-field is required, if multiple logical channels are mapped on the same DCH-TrCH.
⇒ When DTCH or DCCH are mapped on the shared DSCH-TrCH, the allocation of a DSCH-RNTI is required. As the figure illustrates, the MAC-header for DCCH/DTCH on a DSCH consists of the identifier of the DSCH-RNTI, the DSCH-RNTI itself and, optionally, the C/T-field, if multiple logical channels are mapped on this DSCH.
⇒ When DTCH or DCCH are mapped on a common CPCH-TrCH, the MAC-header consists of the identifier of the C-RNTI, the C-RNTI itself and, optionally, the C/T-field, if multiple logical channels are mapped on this CPCH.
⇒ When DTCH or DTCH are mapped on a common RACH-TrCH, the MAC-header consists of the TCTF-field, the identifier of the C-RNTI, the C-RNTI itself and, mandatorily, the C/T-field, because there are always multiple DCCH-/DTCH-logical channels defined that need to be distinguished per UE.

[3GTS 25.321 (9.2.1)]

The Different RNTI's

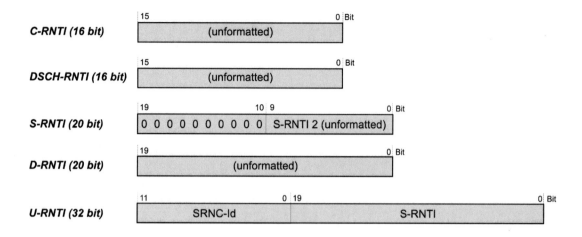

The Different RNTI's

There are 5 different RNTI's:

C-RNTI
The C-RNTI is allocated by the CRNC and uniquely identifies a UE within one cell. The existence of the UE at a given moment in time is optional. However, if common transport channels (FACH, CPCH, RACH) shall be used as bearer for DCCH- / DTCH-logical channels, the allocation of the C-RNTI is mandatory. This applies in particular in CELL_FACH-state.

DSCH-RNTI
The DSCH-RNTI is allocated by the CRNC when DSCH-TrCH's shall be used as bearer for DCCH / DTCH and uniquely identifies a UE within one cell.

S-RNTI
The S-RNTI is allocated by the SRNC upon RRC-connection establishment and uniquely identifies a UE within the SRNS. In a macrodiversity situation, the S-RNTI is used by the DRNC to identify the UE towards the SRNC within RNSAP-messages. The S-RNTI becomes invalid upon RRC-connection release and when the SRNC changes ($\Leftrightarrow$ new S-RNTI during SRNS-relocation). Note that only the bits 0 – 9 of the S-RNTI contain useful information ($\Leftrightarrow$ S-RNTI 2) while the bits 10 – 19 are fixed coded with '0'.

D-RNTI
The D-RNTI (Drift) is allocated by the CRNC/DRNC upon a UE accessing a cell which is controlled by this RNC or when a radio link shall be setup for this UE within the DRNC ($\Leftrightarrow$ macrodiversity). The D-RNTI is provided to the SRNC and serves as identifier for the UE in RNSAP-messages from SRNC to DRNC.

U-RNTI
The U-RNTI consists of the S-RNTI and the SRNC-Id. Therefore, the allocation of the S-RNTI really means that the U-RNTI is allocated. The U-RNTI is allocated to a UE upon RRC-connection establishment and released upon RRC-connection release. It needs to be re-allocated upon SRNC-relocation. Since the U-RNTI also includes the SRNC-Id, it will uniquely identify the UE within the UTRAN.

> Note: The D-RNTI is used on RNSAP-level between DRNC and SRNC on the Iur-interface to identify the UE. The D-RNTI is never used on the Uu-interface. All other RNTI's are allocated by the CRNC or by the SRNC through the RRC-protocol. These other RNTI's are used within the MAC-protocol and the RRC-protocol.

[3GTS 25.401 (6.1.7), 3GTS 25.301 (6)]

Example: Allocation of U-RNTI to UE

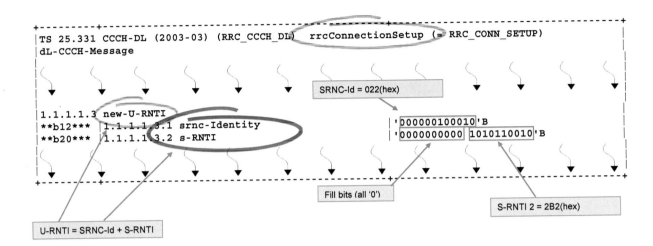

Example: Allocation of U-RNTI to UE

The example illustrates part of an RRC_CONN_SETUP-message which is received by a UE. This message will mandatorily convey the new allocated U-RNTI to the UE.

UMTS – Signaling & Protocol Analysis

Practical Exercise:

C-RNTI → Identity of an MS

- The following strings represent MAC-PDU's which have been sent or received by the UE on different TrCH's. Please identify and determine the different fields of the respective MAC-headers:

TCTF = Target channel Type field.

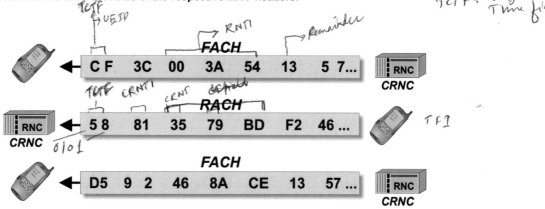

TFI

Answer:

UMTS – Signaling & Protocol Analysis

Answer:

The MAC-Architecture

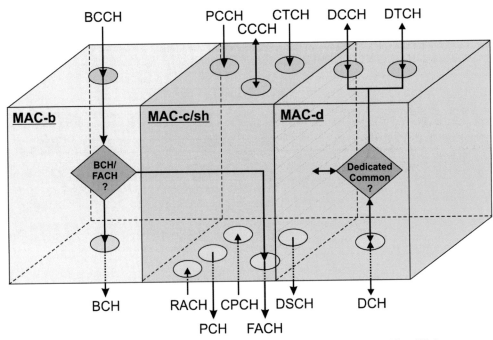

The MAC-Architecture

The MAC-protocol is actually split into 3 different functional units. The split is due to the responsibility for the different transport channels and different logical channels:

- **MAC-b (Broadcast)**
  MAC-b takes care of the administration of the BCCH logical channel and the BCH-TrCH. On the network side, MAC-b is physically located in the NodeB. In the UE, there are multiple MAC-b-entities; one for the serving cell and possibly several others for the different neighbor cells. As the figure illustrates, there is the possibility to send BCCH-information through the FACH-TrCH but this only applies for SIB 10 and only for UE's that support the simultaneous reception of one ore more DPDCH/D and one S-CCPCH.

- **MAC-c/sh (common / shared)**
  MAC-c/sh takes care of the administration of all common logical channels (PCCH, CCCH and CTCH) and all common and shared transport channels (RACH, CPCH, DSCH, FACH, PCH). On the network side, MAC-c/sh is physically located in the CRNC (*not in the SRNC!*). There is one MAC-c/sh-entity in the CRNC for each cell that is connected to it. In the UE, there is only one MAC-c/sh-entity. Common logical channels can only be mapped on common and shared transport channels.

- **MAC-d (dedicated)**
  MAC-d takes care of the administration of the dedicated logical channels (DCCH and DTCH) and of the dedicated transport channels (DCH). On the network side, MAC-d is physically located in the SRNC. There is one MAC-d-entity in the SRNC for each UE that has currently established dedicated logical channels to this SRNC. Note that dedicated logical channels can be mapped on common or shared transport channels. The configuration is taken care of by RRC but the decision is QoS- and load-dependent.

[3GTS 25.321 (4.2)]

MAC-b

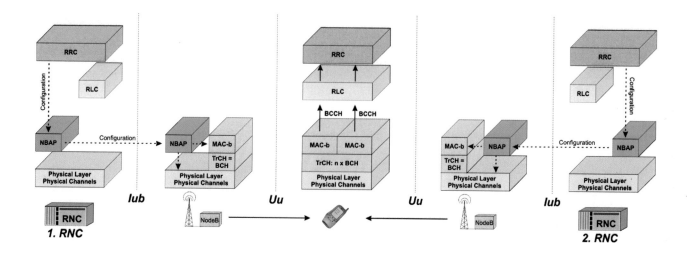

MAC-b

⇒ The MAC-b sublayer (⇔ abbreviation for "broadcast") is in charge for the BCH-TrCH. As the figure illustrates, there may be multiple instances of the MAC-b sublayer and therefore of the BCH in the UE (see note underneath).
⇒ On the network side, the MAC-b sublayer is logically and physically integral part of the NodeB. For each cell, the NodeB requires a separate MAC-b entity.
⇒ On the network side, the MAC-b sublayer cannot be configured directly by the RRC-protocol. This configuration and therefore also the specification of the content of the various system information messages is a task of the NBAP-protocol. Obviously, NBAP acts only as a relay protocol between RRC and MAC.

Note:
- Each UE shall be capable to support up to 32 FDD-neighbor cells on the same frequency plus 32 more FDD-neighbor cells on maximum 2 more frequencies.
- The additional support of 32 TDD-neighbor cells on up to 3 different frequencies and / or 32 GSM-cells on up to 32 different frequencies is optional and depending on UE-capabilities.

[3GTS 25.321 (4.2.2), 3GTS 25.133 (8.1.2.1)]

MAC-c / sh

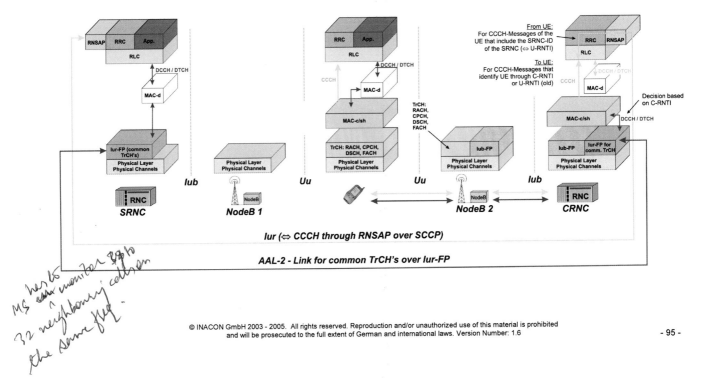

MAC-c / sh

⇒ The MAC-c/sh sublayer (⇔ abbreviation for "common / shared") is in charge for all common transport channels (incl. shared transport channels like the DSCH). As the figure illustrates, there may be only one instance of the MAC-c/sh sublayer in the UE.
⇒ On the network side, the MAC-c/sh sublayer is logically and physically integral part of the Controlling RNC (CRNC). There is one MAC-c/sh entity for each cell in a CRNC.

> The figure illustrates a situation when the SRNC is different from the CRNC. For MAC-c/sh, this is a considerable situation *only* after a cell update scenario towards another RNC without SRNC-relocation.

⇒ In such a case, all common transport channels will not terminate in the SRNC but in the CRNC.
⇒ If the CRNC receives an RRC-message on CCCH from the UE, it will be able to identify the SRNC based on the SRNC-Id which is part of the U-RNTI which is definitely included in the RRC-message. The respective signaling message can be a CELL_UPD-message or a URA_UPD-message (⇔ RRC-chapter). These signaling messages will be forwarded to the SRNC by the RNSAP-protocol (⇔ Uplink Signaling Transfer Procedure) over the Iur-interface. The CRNC will allocate a C-RNTI and a D-RNTI for the UE and inform the SRNC about these RNTI's as part of this procedure.
⇒ Vice versa, signaling messages on CCCH from the SRNC to the UE and through the CRNC will also be sent by the RNSAP-protocol (⇔ Downlink Signaling Transfer Procedure) over the Iur-interface. The initial RRC-message on CCCH towards the UE will include the U-RNTI and the C-RNTI of the UE.
⇒ If the SRNC does not decide for an SRNC-relocation and DCCH- or DTCH-information need to be relayed to the UE via the CRNC, then an AAL-2 link for the transfer of common/shared transport channels over the Iur-interface needs to be established between SRNC and DRNC (⇔ RNSAP / Common Transport Channel Resources Initialization Procedure). An example is illustrated in the chapter "Important UMTS-Scenarios".

Note:
- Macrodiversity is not defined for common nor shared transport channels (RACH, CPCH, FACH, DSCH). Still, common/shared channel data streams can be transferred over the Iur-interface through the Iur-frame protocol for common shared channels [3GTS 25.425]. An AAL-2 link is required between SRNC and CRNC to transfer the relates Iur-FP PDU's.
- If there is already an AAL-2 link between the SRNC and the CRNC for common transport channels (for other UE's), then the SRNC will not request the establishment of another AAL-2 link for common/shared transport channels.

[3GTS 25.321 (4.2.3)]

MAC-d

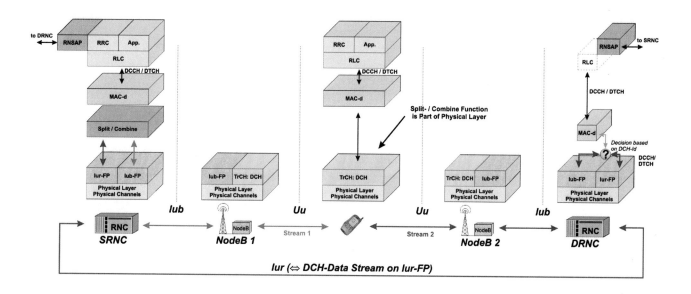

MAC-d

⇒ The MAC-d sublayer (⇔ abbreviation for "dedicated") is in charge for all dedicated transport channels. As the figure illustrates, there may be only one instance of the MAC-d sublayer in the UE.
⇒ On the network side, the MAC-d sublayer is logically and physically integral part of the Serving RNC (SRNC). There is exactly one MAC-d entity in the SRNC for every UE that has one or more DCH's setup towards the SRNC.
⇒ The figure illustrates a situation when there is an SRNC and a CRNC. In the illustrated case, there are 2 DCH-data streams between the UE and the SRNC: The first stream (⇔ Stream 1) is routed through NodeB 1 which is connected to the SRNC. The second stream (⇔ Stream 2) is routed through NodeB 2 which is connected to its CRNC which is the DRNC for this specific macrodiversity situation.
⇒ The SRNC has configured the second stream through the DRNC and NodeB 2 after receiving a measurement report from the UE, indicating that a cell of the neighbor NodeB 2 can provide a better service than the currently serving cell which is part of NodeB 1.
⇒ To provide for the second DCH-data stream through NodeB 2 and the DRNC, there is an additional DCH-data stream required between SRNC and DRNC over the Iur-interface.
⇒ The DRNC will deploy a fixed mapping between Iub-FP data streams that have to delivered to the local MAC-d entity and those Iub-FP data streams that have to be forwarded on the Iur-interface towards the SRNC.
⇒ In downlink direction, the mobile station will experience the two (or more) different DCH-data streams as different radio links with different or equal channelization codes but different scrambling codes of the same signal. The combination of the two or more data streams already occurs in the physical layer through soft decisions.
⇒ In uplink direction, the SRNC requires a specific split/combine-function underneath MAC-d to decide, how the two or more uplink DCH-data streams shall be processed by MAC-d and upwards to higher layers.

Note:
- The split-/combine function in the SRNC receives the already de-interleaved and channel decoded plain information bits from the DRNC.
- Still, based on the QE-information (Quality Estimate) delivered in uplink data frames of the Iur-FP and Iub-FP [3GTS 25.427 (6.2.4.5)] the split/combine function is able to weigh the possibly several DCH-data streams differently and to combine them according to this weighting.
- The QE either contains the measured bit error rate (BER) for the uplink physical control channel (⇔ default / based on information bits on DPCCH/U) or the BER of the information bits within the DPDCH's of one selected TrCH. The selection is done by the RNC upon radio link setup.
- The QE-value ranks from '0' to '255'. While the value '0' represents (0% ≤ BER ≤ 0.863475%) the value 255 represents BER = 100% (see also RRC-section / "Important Measurement Parameters".
- Similarly, the RNC may request the UE to conduct and report quality measurements that perform the same evaluation procedures in downlink direction.

[3GTS 25.321 (4.2.4)]

Traffic Volume Measurements and Radio Bearer Control

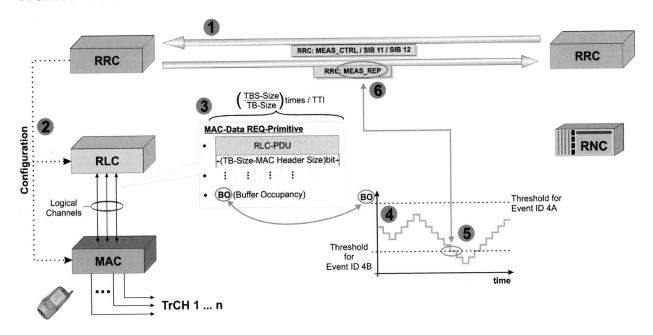

Traffic Volume Measurements and Radio Bearer Control

The figure illustrates it: The RNC *may* configure traffic volume measurements for uplink TrCH's in the UE. This configuration relates to the definition of minimum and maximum thresholds for the BO (Buffer Occupancy). Consecutively, the UE will convey measurement reports to the RNC, either event triggered or periodically.

Two event ID's are defined:
⇒ Event ID 4A: The BO exceeds the maximum threshold.
⇒ Event ID 4B: The BO becomes smaller than the minimum threshold.
As a consequence of these traffic volume measurements, the RNC may reconfigure the radio bearer (⇔ TrCH's) to cope with the new requirements.

Note: The RNC may also configure traffic volume measurements for downlink TrCH's in the local MAC-entity.

Let us evaluate in more detail how traffic volume measurements are performed:
1. The RNC has different options to activate traffic volume measurements within the UE: a) through SIB 11 b) through SIB 12 (optional and only in RRC-connected mode) and c) by transmitting a dedicated RRC: MEAS_CTRL-message to the UE that specifically configures traffic volume measurements and the way how the UE shall convey the result to the RNC (event-triggered or periodically).
2. The RRC-protocol within the UE will react upon the received setup information and send configuration primitives to RLC and MAC.
3. The third bullet highlights a singe MAC-Data REQ-primitive that is sent from RLC to MAC to request the transfer of an RLC-PDU-segment. Please note that among other things, this primitive contains the current BO) of this logical channel. BO relates to the entire data volume that is currently buffered in RLC for this logical channel. Such MAC-Data REQ-primitives are sent n times per TTI from RLC to MAC whereas "n" is defined through the number of transport blocks (TB) per transport block set (TBS).
4. The MAC-layer will keep track over time how the aggregated BO for the different TrCH's changes. The diagram illustrates such a situation in which at point 5
5. the predefined threshold for reporting event 4B is reached.
6. This event triggers the transfer of a traffic volume measurement report towards the RNC. How and when the RNC reacts upon this measurement report is implementation dependent.

[3GTS 25.321 (11.1), 3GTS 25.331 (14.4)]

The Random Access Procedure

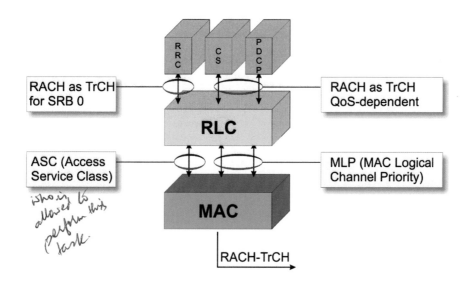

The Random Access Procedure

The random access procedure always needs to be executed when the RACH-TrCH shall be used as transport channel for higher layer data.

Initial Conditions
The figure illustrates when the RACH-TrCH is used:

⇒ When the RRC-protocol needs to send an RRC-message over the SRB 0 (Signaling Radio Bearer).
⇒ When CS-services (circuit-switched) or packet-switched services (PDCP) have previously been configured during a Radio Bearer Setup-procedure to use the RACH-TrCH.

> **Note:**
> - On RACH-TrCH, only short information packages can be sent. The absolute length depends on the configured transport format (⇔ SIB5).
> - Only the less sophisticated open loop power control procedure is applicable for the RACH (⇒ interference risk).
> - Therefore, the use of RACH-TrCH for circuit-switched or packet-switched data transfer should only be considered for applications with very robust delay and reliability requirements (or, like RRC, the application deploys its own reliability scheme).

[3GTS 25.321 (11.2)]

Overview

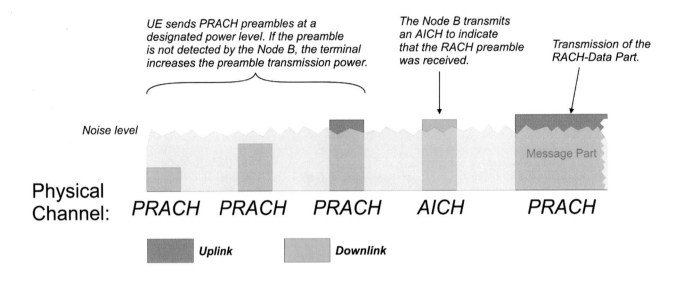

Overview

The RACH-TrCH is a common transport channel that uses the PRACH as physical channel. Multiple UE's need to share the available resources. The underlying PRACH deploys a slotted aloha access scheme with fast acquisition indication to grant resources to a requesting UE. We will elaborate the details on the following pages. This page shall only provide an overview:

- During the slotted aloha phase, the UE will transmit PRACH-preambles in well defined RACH access slots on the PRACH.
- Each PRACH-preamble contains 256 repetitions of one out of 16 different PRACH-signatures.
- The SIB 5 defines which signatures are available in a cell.
- Note that the UE will randomly re-select one of the available signatures for each preamble transmission.
- Whether or not a UE is allowed to transmit a preamble is decided by the relative priority of the network access. This priority is called ASC (access service class).
- Each preamble is transmitted with a higher output power. Eventually, a preamble will exceed the current noise level in the cell and will be received by the NodeB.
- Consequently, the NodeB will positively (or negatively ;-) acknowledge reception of the preamble by mirroring the PRACH-signature back to the UE on the AICH.
- Reception of its last PRACH-signature at a pre-defined time on the AICH tells the UE that it is now allowed to transmit the PRACH-message part.

Access Service Class Selection

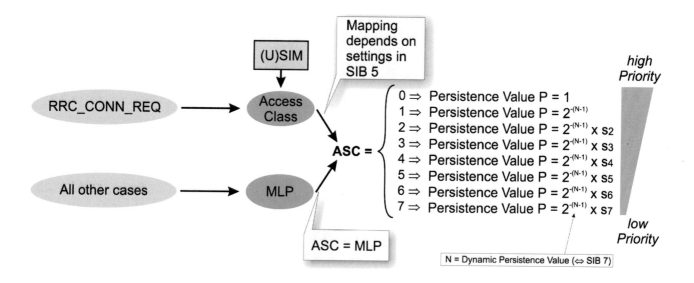

Access Service Class Selection

During the slotted aloha phase, the preamble transmissions from different UE's may collide. To give UE's with a higher access reason a better chance to get through on the PRACH, UTRAN uses different persistence levels P for those UE's. These persistence levels are mostly controlled by UTRAN itself and provide means to decide, whether or not a UE may transmit an access preamble.

Transfer of an RRC_CONN_REQ-Message
If the network access is due to the transmission of an RRC_CONN_REQ-message (initial access), the RRC-layer will map the AC (access class) from the SIM/USIM to the UMTS-specific ASC (access service class = 0 ... 7). This possibility is applicable only if the UE intends to switch from RRC-Idle-state to RRC-connected state. The mapping between AC and ASC is provided per cell in SIB 5.

Other Cases (MLP)
If the network access is happening after radio bearers have been established already, the MLP-parameter (MAC logical channel priority) is used to determine the relative priority for the network access. The RNC will allocate an MLP to every logical channel that is being configured during the RRC connection setup or radio bearer setup/re-configuration procedure (see for example in RRC_CONN_SETUP-messages or RB_SETUP-messages). Although MLP can be in the range between '1' and '8', the lowest possible ASC = 7, even when MLP = 8.

Determination of the Persistence Value P
Having found the correct ASC, the determination of the persistence value P also depends on the fast changing "dynamic persistence value" N (⇔ broadcast in SIB 7 / value range = 1 .. 8) and the "persistence scaling factors" Si (⇔ optionally broadcast in SIB 5 / 6). In general, the persistence value P will be higher with a lower ASC but this is no absolute rule because of the "persistence scaling factors" Si. The highest persistence value is P = 1 which relates to ASC = 0. ASC = 0 should be reserved for emergency call. The higher the value of the "dynamic persistence value" N is selected, the lower the possibility becomes for the other ASC's 1 to 7 to transmit a PRACH-preamble.

> **Note:**
> - Through the adjustment of the "dynamic persistence value" N and the "persistence scaling factors" Si the NodeB can dynamically increase or decrease the chances for each ASC to transmit a PRACH-preamble.

[3GTS 25.321 (11.2), 3GTS 25.331 (8.5.12)]

Persistence Value P and Random Number R

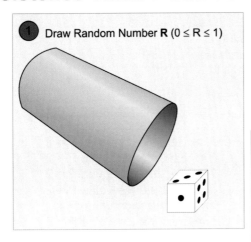
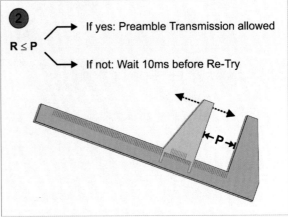

Persistence Value P and Random Number R

After the calculation of the persistence value P the following procedure is used to decide whether or not the transmission of an access preamble is legitimate:

1. The UE shall draw a random number R in the range $0 \leq R \leq 1$.
2. If $R \leq P$, then the transfer of a PRACH-preamble is allowed (if all other conditions apply). Otherwise, wait 10 ms before retry.

- **Example 1: P = 0.9**
  Accordingly, the probability to send a PRACH-preamble is 90%. Statistically, in 9 out of 10 cases, the UE can send a PRACH-preamble. Another point of view: The transmission of 10 PRACH-preambles would take app. 110 ms.

- **Example 2: P = 0.5**
  Accordingly, the probability to send a PRACH-preamble is 50%. Statistically, in 5 out of 10 cases, the UE can send a PRACH-preamble. Another point of view: The transmission of 10 PRACH-preambles would take app. 200 ms.

- **Example 3: P = 0.1**
  Accordingly, the probability to send a PRACH-preamble is 10%. Statistically, in 1 out of 10 cases, the UE can send a PRACH-preamble. Another point of view: The transmission of 10 PRACH-preambles would take app. 1 s.

[3GTS 25.321 (8.5.12)]

PRACH Access Slots and AICH Access Slots

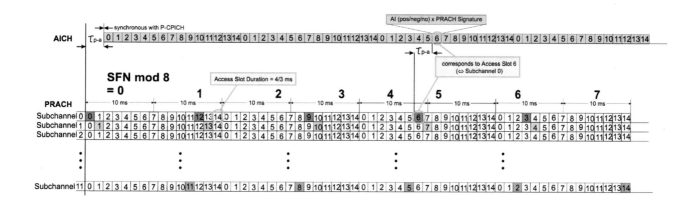

PRACH Access Slots and AICH Access Slots

There may be more than one PRACH physical channel defined per cell (different channelization codes) and each PRACH is further divided into 12 PRACH subchannels. Which PRACH's are available and which subchannels on each PRACH are available, is broadcast in SIB 5

PRACH-Subchannels
⇒ The 12 different PRACH-subchannels are illustrated in the figure.
⇒ Each subchannel consists of 5 PRACH-access slots over a period of 80 ms or 8 radio frames.
⇒ Accordingly, 60 PRACH-access slots are included in these 80 ms. Each access slot has a duration of 4/3 ms.

Which subchannels can be used by a UE is determined by this UE through IE's within SIB 5 (RRC-idle and connected mode) or SIB 6 (RRC-connected mode only).

Selection of an Appropriate PRACH-Access Slot
⇒ Whenever a UE intends to transmit a PRACH-access preamble, it shall within a 10 ms radio frame randomly select one available PRACH-subchannel which also tells the UE the very PRACH-access slot that can be used.
⇒ Consequently, the UE will apply the persistence calculation based on the ASC, as presented on the previous page. If R ≤ P (with R = Random Value and P = Persistence Value), the UE will transmit an access preamble. Otherwise, the UE has to wait 10 ms (⇔ 1 radio frame) before the next PRACH-access slot selection can take place.

Response from the NodeB on AICH / Time Relation
⇒ The time relation between the PRACH-access slots and the AICH is illustrated in the figure: The PRACH proceeds the AICH by τ_{p-a}. The time value τ_{p-a} depends on the IE "AICH Transmission Timing" in SIB 5 / 6: τ_{p-a} = 7860 chips or 12800 chips (3 radio frames or 5 radio frames). The AICH is time aligned with the primary CPICH.
⇒ As the figure illustrates, the UE will receive an acquisition indication (positive, negative, none) on AICH exactly τ_{p-a} after the start of the access preamble transmission.

[3GTS 25.214 (6.1), 3GTS 25.321 (11.2.2)]

Preambles, Acquisition Indication and Message Part Transfer

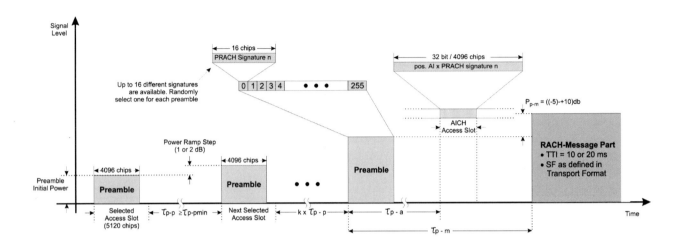

Preambles, Acquisition Indication and Message Part Transfer

The figure illustrates the transmission of the PRACH access preambles, the acquisition indication and the RACH-message part in conjunction with the related timing and power constraints:

⇒ The UE will randomly select one PRACH signature (according to SIB 5) and one PRACH subchannel. Then the UE will convey the initial PRACH access preamble with "Preamble Initial Power" and wait for an acquisition indication on AICH τ_{p-a} chips after the start of the access preamble transmission.

⇒ If there is no positive nor negative acquisition indication received by the UE, the UE will randomly select another signature and another subchannel number. This new preamble is transmitted with a power level that is "Power Ramp Step" higher than the previous transmission.

⇒ Eventually and definitely before the maximum number of preamble transmissions has been performed, the UE needs to receive a positive acquisition indication on AICH.

⇒ As can be seen, this positive acquisition indication needs to be received exactly τ_{p-a} chips after the start of the access preamble transmission.

⇒ Having received the positive acquisition indication, the UE will start transmitting the RACH-message part exactly τ_{p-m} chips after the start of the access preamble transmission. Note that the power level between the final access preamble and the RACH-message part is controlled by the parameter Pp-m.

⇒ The length of RACH-message part (10 ms or 20 ms) is like the transport format set for the RACH defined in SIB 5.

[3GTS 25.214 (6.1), 3GTS 25.321 (11.2.2)]

RACH / PRACH Configuration in SIB 5 (Part 1)

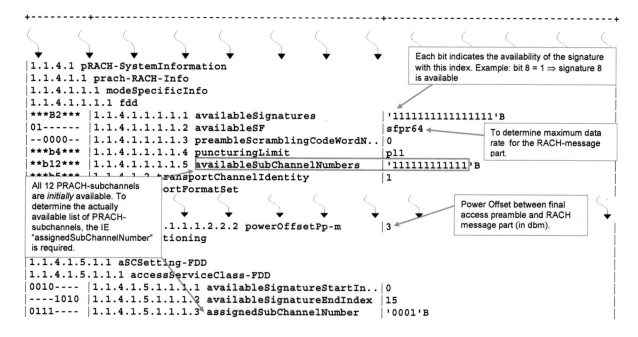

RACH / PRACH Configuration in SIB 5 (Part 1)

- Determination of the available PRACH subchannels with AICH Transmission Timing = 0

  ```
  1111 1111 1111                ⇐ availableSubChannelNumbers
  AND
  0010 0100 1001                ⇐ 4 times the 3 least significant bits of the assignedSubChannelNumber
  0010 0100 1001                ⇐ Available Subchannels are subchannel 0, 3, 6 and 9
  ```

- Determination of the available PRACH subchannels with AICH Transmission Timing = 1

  ```
  1111 1111 1111                ⇐ availableSubChannelNumbers
  AND
  0001 0001 0001                ⇐ 3 times the entire assignedSubChannelNumber
  0001 0001 0001                ⇐ Available Subchannels are subchannel 0, 4 and 8
  ```

[3GTS 25.331 (8.6.6.29)]

RACH / PRACH Configuration in SIB 5 (Part 2)

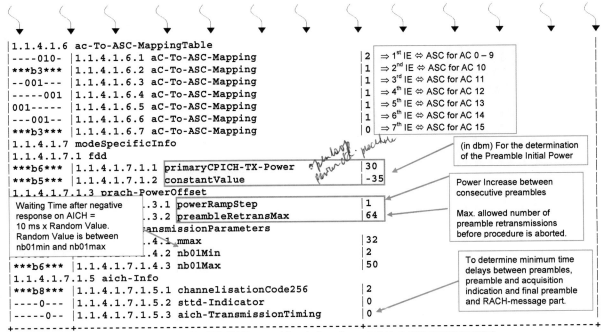

RACH / PRACH Configuration in SIB 5 (Part 2)

- **Preamble Initial Power**

  > Preamble Initial Power = Primary CPICH TX Power − CPICH RSCP + UL Interference + Constant Value.

  ⇒ CPICH RSCP ⇔ to be measured by the UE
  ⇒ UL Interference ⇔ SIB 7

- **AICH Transmission Timing**
  First value: AICH Transmission Timing = 0 / Second value: AICH Transmission Timing = 1

  ⇒ $\tau_{p\text{-}p}(min)$ = 15360 chips / 20480 chips
  ⇒ $\tau_{p\text{-}a}$ = 7860 chips / 12800 chips
  ⇒ $\tau_{p\text{-}m}$ = 15360 chips / 20480 chips

  [3GTS 25.331 (8.6.6.29, 8.5.12, 8.5.13)]

Radio Link Control (RLC)

- **Operation Modes**

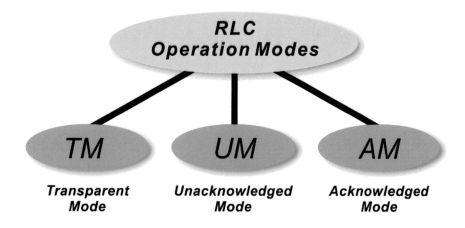

Radio Link Control (RLC)

Operation Modes

Transparent Mode (TM)
In transparent mode, the RLC-protocol will simply convey higher layer SDU's to the peer RLC-entity but RLC will neither provide error detection and correction mechanisms nor ciphering nor can RLC guarantee in-sequence delivery. Segmentation of RLC-SDU's is possible but all segments of one RLC-SDU must fit into the RLC-TMD-PDU's that are forwarded to MAC within a single TTI. No padding is allowed. Accordingly, the TM is only applicable, if the application knows exactly the size of the RLC-SDU's to be transferred. RLC will add no header information to an RLC-TMD-PDU.

> The RLC transparent mode is most applicable for the transfer of real-time service related PDU's of the same size between UE and RNC (e.g. AMR-speech frames). The time relation is more important than whether all packets arrive and whether these packets arrive in sequence.

Unacknowledged Mode (UM)
In unacknowledged mode, the RLC-protocol will also convey higher layer SDU's to its peer. These SDU's are segmented by RLC into RLC-UMD-PDU's and a header is appended to each segment. Part of this header is a sequence number (modulo 7). If one RLC-UMD-PDU is received out of sequence, the entire related RLC-SDU will be discarded. Unlike in TM, the size of RLC-SDU's for UM cannot necessarily be predicted by the application. Accordingly, the UM provides for concatenation of RLC-SDU'-segments into a single RLC-UMD-PDU and for padding within RLC-UMD-PDU's.

> The RLC unacknowledged mode is most applicable for higher layer data that require time consistent (⇔ almost real-time) and in-sequence data transfer between UE and RNC (e.g. web-surfing). Error correction mechanisms need to be taken care of by higher layers.

Acknowledged Mode (AM)
In acknowledged operation mode, RLC guarantees the error-free and in-sequence delivery of higher layer SDU's to the peer. Obviously, retransmissions may be required and therefore, the time relation cannot be guaranteed in AM. In addition, the AM provides for flow control through variable window size. The related timers and counters need to be configured between sender and receiver prior to any SDU-transfer. This configuration is taken care of by the RNC during bearer setup (e.g. RRC_CONN_SETUP-message, RB_SETUP-message).

> The RLC acknowledged operation mode is most applicable for error sensitive higher layer data when the time consistency is of less importance (e.g. file transfer, transfer of signaling messages on DCCH). The AM operation mode is only applicable on DCCH (⇔ control plane) and DTCH (⇔ user plane).

[3GTS 23.322 (6, 11.1, 11.2, 11.3)]

Tasks and Functions

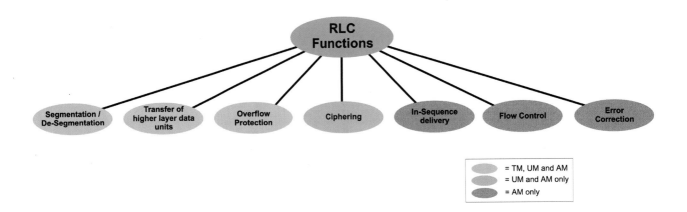

Tasks and Functions

- **Segmentation and De-Segmentation of Higher Layer PDU's**
  The way how RLC segments the higher layer PDU's into RLC-PDU's depends mostly on the configured TB-sizes for the related TrCH. Additionally, for TM, all segments need to be identically sized and each segment needs to fit into one RLC-TMD-PDU.

- **Transfer of Higher Layer PDU's**
  As a classical data link layer (layer 2) implementation, the main function of RLC is the exchange of higher layer PDU's between RNC and UE.

- **Overflow Protection**
  Depending on the RLC-operation mode (TM, UM, AM), the overflow protection may simply consist of SDU-discard in case of overload or of more sophisticated mechanisms like SDU-discard after a maximum number of retransmissions has been performed.

- **Ciphering**
  The encryption of the content of RLC-PDU's is performed for RLC-UMD-PDU's and RLC-AMD-PDU's, if configured by RRC.

- **In Sequence-Delivery**
  In-sequence delivery is only applicable for AM. The numbering of RLC-UMD-PDU is performed to detect out of sequence delivery. However, in such a case the related SDU will be discarded and therefore, UM cannot guarantee in-sequence delivery.

- **Flow Control**
  Flow control relates to an increased or decreased SDU-transmission rate through the adjustment of the window sizes. The larger the window size, the smaller the round trip time.

- **Error Correction**
  MAC will inform RLC for each received RLC-PDU whether this RLC-PDU is garbled (⇔ MAC-DATA-Indication primitive / Error Information, based on the CRC-check in physical layer). In such cases, it is the responsibility of RLC to request the retransmission of this RLC-PDU from the peer.

Configuration of the Size of an RLC-PDU

```
+----------+----------------------------------------------+------------------+
|BITMASK   |ID Name                                       |Comment or Value  |
+----------+----------------------------------------------+------------------+
|RRC_CONN_SETUP                                            |
|
|
|1.1.1.1.9 ul-AddReconfTransChInfoList
|1.1.1.1.9.1 uL-AddReconfTransChInformation
|-----0--  |1.1.1.1.9.1.1 ul-TransportChannelType         |dch
|***b5***  |1.1.1.1.9.1.2 transportChannelIdentity        |32
|1.1.1.1.9.1.3 transportFormatSet
|1.1.1.1.9.1.3.1 dedicatedTransChTFS
|1.1.1.1.9.1.3.1.1 tti
|1.1.1.1.9.1.3.1.1.1 tti40
|1.1.1.1.9.1.3.1.1.1.1 dedicatedDynamicTF-Info
|1.1.1.1.9.1.3.1.1.1.1.1 rlc-Size
|1.1.1.1.9.1.3.1.1.1.1.1.1 octetModeType1
|***b5***  |1.1.1.1.9.1.3.1.1.1.1.1.1.1 sizeType1         |16
|
+----------+----------------------------------------------+------------------+
```

The actual RLC-PDU Size for this DCH/UL is provided in octetmodeType 1:
⇒ RLC-PDU-Size = (8 x sizeType1) + 16 [bit]
⇒ RLC-PDU-Size = (8 x 16) + 16 = **144 bit**

Extract of 3GTS 25.331 (ASN.1-Code):

```
OctetModeRLC-SizeInfoType1 ::=         CHOICE {
-- Actual size = (8 * sizeType1) + 16
    sizeType1                      INTEGER (0..31),
    sizeType2                      SEQUENCE {
    -- Actual size = (32 * part1) + 272 + (part2 * 8)
        part1                      INTEGER (0..23),
        part2                      INTEGER (1..3)
    },
    sizeType3                      SEQUENCE {
-- Actual size = (64 * part1) + 1040 + (part2 * 8)
        part1                      INTEGER (0..61),
        part2                      INTEGER (1..7)
    }
}
```

Configuration of the Size of an RLC-PDU

The extract illustrates how the UE is informed about the configured TB-size. This happens indirectly through the configuration of the related RLC-PDU-size. The RRC-protocol has different options on how to convey the RLC-PDU-size to its peer in the UE:

- **Octet Mode Type 1, 2 or 3**
  The octet mode is applicable, if the TB-sizes shall only vary by at least 8 bit. The different types (⇔ sizeType1, 2, 3) allow for the definition of larger or smaller RLC-PDU-sizes by still using the minimum number of bits to identify these sizes. This is achieved through offsets in sizeType2 and sizeType3.

- **Bit Mode Type 1, 2, 3 and 4.**
  The bit mode is applicable, if the TB-sizes shall vary in bits.

> **Note the Differences between NBAP (Size Configuration RNC ⇔ NodeB) and RRC (Size Configuration RNC ⇔ UE):**
> - In NBAP, the TB-size is conveyed to the NodeB in plain text (e.g. 360 bit). Considering that RLC-PDU-size = TB-Size − MAC-header size, the NodeB needs to know which logical channel ⇔ transport channel mapping is used (because of the different MAC-header sizes) to determine the RLC-PDU-size.
> - The UE needs to approach the problem from the other end. It obtains from the RNC the RLC-PDU-size and the mapping information between logical channels and transport channels. Applying the formula from above, the TB-size = RLC-PDU-size + MAC-header size.

[3GTS 25.331]

RLC-PDU-Types

- ### The TMD-PDU and the UMD-PDU

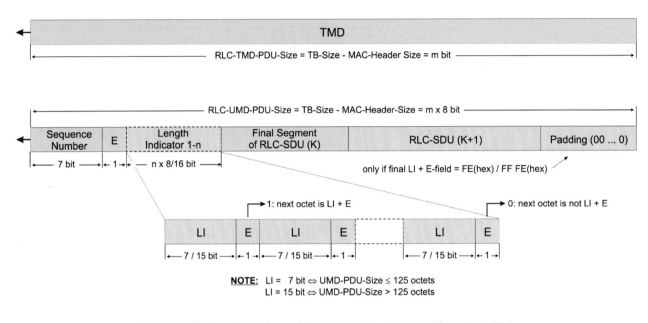

RLC-PDU-Types

The TMD-PDU
Most importantly, the TMD-PDU does not include nor require any RLC-header. The TMD-PDU is m bit long whereas m does not need to be a multiple of 8 bit. There are no optional fields in a TMD-PDU.

> Note: The size of an RLC-PDU is configured by RRC in e.g an RRC_CONN_SETUP- or SIB-message.

The UMD-PDU
The UMD-PDU includes a header with a minimum length of 1 octet. Length indicators within an UMD-PDU are only required, if segments from different RLC-SDU's need to be delimited from each other or if padding is used to fill an UMD-PDU to the configured size.
Unlike the TMD-PDU, the length of a UMD-PDU shall be a multiple of 8 bit. In other words, the UMD-PDU will always terminate on an octet boundary.

- ### Header Fields

    ### Sequence Number (7 bit)
    The sequence number is modulo 7. It shall be incremented for every new UMD-PDU which is transferred from an UM-sender to an UM-receiver. The initial sequence number for UMD-PDU's from a UE to UTRAN shall be '0'. However, the initial sequence in the direction UTRAN ⇒ UE cannot be predicted.

    ### E-Bit
    The E-bit or Extension-bit is used to identify whether or not the next octet contains data or a length indicator.

    ### Length Indicator (7 bit / 15 bit) plus E-Bit
    Length indicators are used to delimit segments from different RLC-SDU's within a single UMD-PDU from each other. In addition, length indicators are used to identify the start of padding within an UMD-PDU.

    > Note: 7 bit length indicators are only applicable as long as the configured UMD-PDU-size is ≤ 125 octets. Since 7 bit can represent values up to 127, this requires some explanation: The limitation 125 is due to the fact, that LI-values = 124, 125, 126 and 127 are reserved or are used for special purposes. Considering a minimum UMD-PDU header length of 1 octet plus another octet for the LI + E-octet, this leaves us with a maximum length indication of 123 octets and therefore a maximum UMD-PDU-length of 125 octets, if 7 bit length indicators shall be used.

[3GTS 25.322 (9.2.1.2, 9.2.1.3, 9.2.2, 11.2)]

Example of an RLC-TMD-PDU

```
+---------+---------------------------------------+---------------------------------------+
|BITMASK  |ID Name                                |Comment or Value                       |
+---------+---------------------------------------+---------------------------------------+
|15:15:41,357,403   2 RACH Cell0  RLC/MAC  FP DATA RACH                                   |
|2 FP:    Transport Block                                                                 |
|00------ |2.1 MAC: Target Channel Type Field     |CCCH (Common Control Channel)  Transparent Mode
|         |2.2 MAC: RLC Mode                      |Transparent Mode                       |
|**b166** |2.3 RLC: Whole Data                    |'0010000100101110000000000000000101'B  |
|         |                                       |1010100010011000100011101100000000'B   |
|         |                                       |0000100001100000000000000000000000'B   |
|         |                                       |0000000000000000000000000000000000'B   |
|         |                                       |0000000000000000000000000000000000'B   |
+----+--+--+--+--+--+--+--+--+--+--+--+--+--+--+--+
|HEX |0 |1 |2 |3 |4 |5 |6 |7 |8 |9 |A |B |C |D |E |F |
+----+--+--+--+--+--+--+--+--+--+--+--+--+--+--+--+
|0   |3c|bb|00|00|08|4b|80|00|ba|89|88|ec|00|20|c0|00|  ← Highlighted octets represent the MAC-content
|10  |00|00|00|00|00|00|00|00|00|00|c3|56|  |  |  |  |
+---------+---------------------------------------+---------------------------------------+
```

Example of an RLC-TMD-PDU

Intentionally left blank

Example of an RLC-UMD-PDU

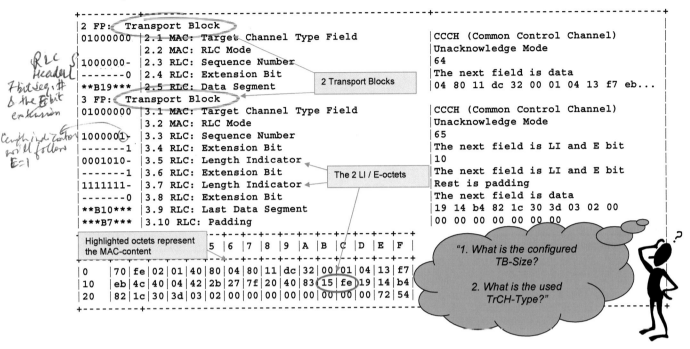

Example of an RLC-UMD-PDU

Answer: 21 TB size

TMD-PDU-Transfer (with segmentation)

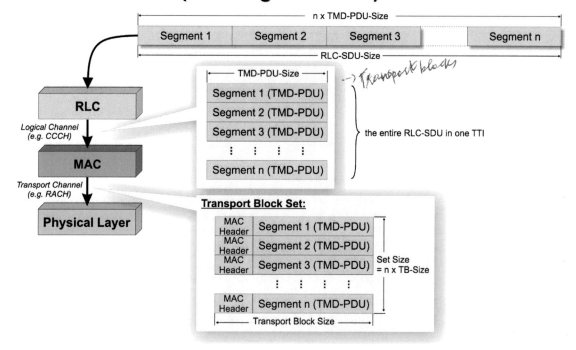

TMD-PDU-Transfer (with segmentation)

The process of transferring RLC-SDU's through RLC-TMD-PDU's is illustrated in the figure. Please recall that only certain *Logical Channel ⇔ Transport Channel combinations* allow for the use of TMD-PDU's.
⇒ In the presented case, RRC has configured that segmentation is applicable in the RLC-layer.

Note: The received RLC-SDU's need to have a size which is a multiple of one of the configured TMD-PDU-sizes.

⇒ The RLC-SDU will be segmented into n segments of equal size (⇔ TMD-PDU-Size). The TMD-PDU-size does not need to be a multiple of 8 bit.
⇒ No RLC-header needs to be added to any of the segments.
⇒ The n segments represent n TMD-PDU's. They are delivered on the configured logical channel to MAC.

Note: The n TMD-PDU's which are delivered to MAC during one TTI represent a single RLC-SDU.

⇒ Depending on the logical/transport channel mapping, MAC may need to append a MAC-header to each TMD-PDU (e.g. if multiple logical channels use the same TrCH). Accordingly, each RLC-PDU is embedded into one MAC-PDU.
⇒ Each MAC-PDU represents a single transport block with a fixed transport block size (per TTI).
⇒ In the next step, MAC will deliver the transport block set to the physical layer for further processing.

Reception of TMD-PDU's: Since there is no numbering nor error detection applied for the TMD-PDU, erroneous or out of sequence TMD-PDU's will not be detected but delivered to the higher layer.

[3GTS 25.322 (4.2.1.1, 11.1)]

TMD-PDU-Transfer (without segmentation)

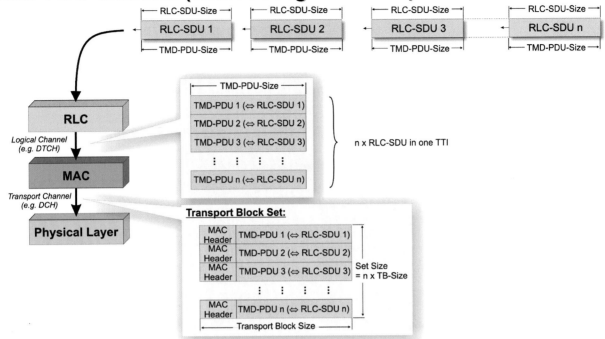

TMD-PDU-Transfer (without segmentation)

⇒ In the presented case, RRC has configured that segmentation is not applicable in the RLC-layer.

Note: The received RLC-SDU's need to have a size which matches *exactly* one of the configured TMD-PDU-sizes. These TMD-PDU-sizes do not need to be a multiple of 8 bit.

⇒ Each RLC-SDU will be mapped into a single TMD-PDU.
⇒ No RLC-header needs to be added to any of the TMD-PDU's.
⇒ The n TMD-PDU's represent n RLC-SDU's. They are delivered on the configured logical channel to MAC.
⇒ Depending on the logical/transport channel mapping, MAC may need to append a MAC-header to each TMD-PDU (e.g. if multiple logical channels use the same TrCH). Accordingly, each RLC-PDU is embedded into one MAC-PDU.
⇒ Each MAC-PDU represents a single transport block with a fixed transport block size (per TTI).
⇒ In the next step, MAC will deliver the transport block set to the physical layer for further processing.

[3GTS 25.322 (4.2.1.1, 11.1)]

UMD-PDU-Transfer

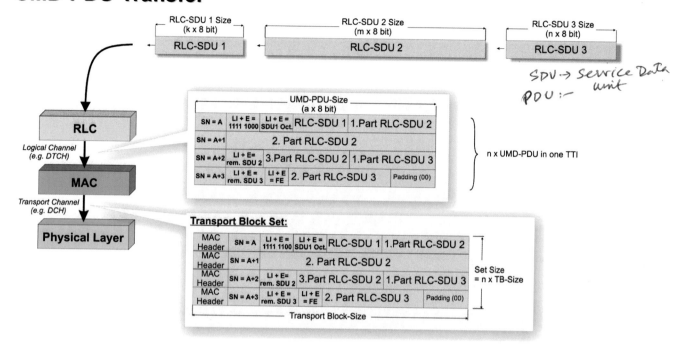

UMD-PDU-Transfer

Note:
- In UM, the RLC-SDU may have a variable size which needs to be a multiple of 8 bit (see figure). Within one TTI, a set of 0, 1 or more UMD-PDU's will be delivered to MAC (⇔ depending on the transport format set). The UMD-PDU-size needs to be a multiple of 8 bit.
- It is not necessary that an entire RLC-SDU is delivered to MAC within one TTI.

⇒ RLC-SDU's will be sliced into possibly several segments of possibly different size. The size of each segment solemnly depends on the available payload space in the UMD-PDU that the respective segment shall be mapped to. However, each segment contains a multiple of 8 bit (the length indicator for delimitation can only point to octet boundaries).
⇒ Each UMD-PDU contains a header that mandatorily contains the sequence number SN of this UMD-PDU and optionally contains more octets for length indicators. In the figure, the LI + E = '1111 1000'$_{bin}$ in SN = A indicates that a new RLC-SDU starts in this UMD-PDU (⇔ downlink only). The second LI + E-octet is required to delimit RLC-SDU 1 from RLC-SDU 2. The LI + E-octet = 'FE'$_{hex}$ in SN = A + 3 indicates that padding is used to fill the UMD-PDU after RLC-SDU 3
⇒ Depending on the size of the RLC-SDU's, the n UMD-PDU's within one TTI represent:
 1. parts of only 1 RLC-SDU or
 2. multiple RLC-SDU's.
⇒ The UMD-PDU's are delivered on the configured logical channel to MAC.
⇒ Depending on the logical/transport channel mapping, MAC may need to append a MAC-header to each UMD-PDU (e.g. if multiple logical channels use the same TrCH). Accordingly, each RLC-PDU is embedded into one MAC-PDU.
⇒ Each MAC-PDU represents a single transport block with a fixed transport block size (per TTI).
⇒ In the next step, MAC will deliver the transport block set to the physical layer for further processing.

Note:
- If the UMD-PDU with SN = A + 2 is received by the peer RLC-entity before UMD-PDU with SN = A + 1 (⇔ out of sequence reception)

or:
- If the UMD-PDU with SN = A + 2 is received by the peer RLC-entity with an error indication (⇔ CRC-check in physical layer failed)

then RLC-SDU's No 2 and No 3 will be discarded by the receiver.

[3GTS 25.322 (4.2.1.2, 11.2)]

The AMD-PDU

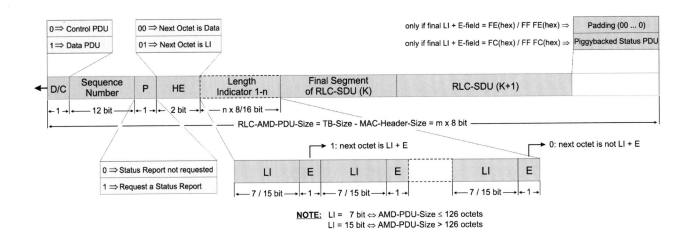

The AMD-PDU

⇒ The AMD-PDU includes a header with a minimum length of 2 octets. Additional header octets for length indicators are only required, if, for instance, segments from different RLC-SDU's are embedded into a single AMD-PDU or if padding is needed to fill up an AMD-PDU to the required size (⇔ as given in the transport format set).
⇒ Like the UMD-PDU, the length of the AMD-PDU shall be a multiple of 8 bit

- **Header Fields**

D/C-Bit
The D/C-bit or Data/Control bit simply distinguishes for AM, whether a AMD-PDU contains data or control information. In case of an AMD-PDU, the D/C-bit = 1.

Sequence Number (12 bit)
The sequence number is modulo 12. It shall be incremented for every new AMD-PDU which is transferred from an AM-sender to an AM-receiver. The initial sequence number for an AM-connection shall be '0' in each direction.

P-Bit
The P-bit or Polling-bit is used to request the receiver of the respective message to convey a status report.

HE-Field (2 bit)
The HE-field or Header Extension-field is used to identify whether or not the next octet contains data or a length indicator.

Length Indicator (7 bit / 15 bit) plus E-Bit
Length indicators are used to delimit segments from different RLC-SDU's within a single AMD-PDU from each other. In addition, length indicators are used to identify the start of padding (⇔ 'FE / FF FE'(hex)) within an AMD-PDU. Yet another LI/E-field value (⇔ 'FC / FF FC'(hex)) is reserved to allow the piggybacking of an RLC-STATUS-PDU into the payload field of an AMD-PDU.

Note: 7 bit length indicators are only applicable as long as the configured AMD-PDU-size is ≤ 126 octets. This is different to the maximum UMD-PDU-length for 7 bit length indicators which was 125 octets. The difference is due to the AMD-PDU-header which is 1 octet longer than the UMD-PDU-header.

[3GTS 25.322 (9.2.1.4, 9.2.2, 11.3)]

Example of an RLC-AMD-PDU

```
+----------+--------------------------------------+-------------------------------------+
|BITMASK   |ID Name                               |Comment or Value                     |
+----------+--------------------------------------+-------------------------------------+
|0010----  |2.1.1  MAC: C/T Field                 |Logical Channel 3                    |
|          |2.1.2  MAC: Target Channel Type       |DCCH (Dedicated Control Channel)     |
|          |2.1.3  MAC: RLC Mode                  |Acknowledge Mode                     |
|----1---  |2.1.4  RLC: Data/Control              |Acknowledged mode data PDU           |
|**b12***  |2.1.5  RLC: Sequence Number           |1                                    |
|-1------  |2.1.6  RLC: Polling Bit               |Request a status report              |
|--01----  |2.1.7  RLC: Header extension type     |Octet contains LI and E bit          |
|***b7***  |2.1.8  RLC: Length Indicator          |9                                    |
|----1---  |2.1.9  RLC: Extension Bit             |The next field is LI and E bit       |
|***b7***  |2.1.10 RLC: Length Indicator          |Rest is padding                      |
|----0---  |2.1.11 RLC: Extension Bit             |The next field is data               |
|**b72***  |2.1.12 RLC: Last Data Segment         |93 02 08 00 08 00 11 82 20           |
|**b40***  |2.1.13 RLC: Padding                   |00 00 00 00 00                       |
+---+--+--+--+--+--+--+--+--+--+--+--+--+--+--+--+
|HEX|0 |1 |2 |3 |4 |5 |6 |7 |8 |9 |A |B |C |D |E |F |
+---+--+--+--+--+--+--+--+--+--+--+--+--+--+--+--+
|0  |7c|14|01|28|00|d1|3f|e9|30|20|80|00|80|01|18|22|
|10 |00|00|00|00|00|00|01|00|82|17|  |  |  |  |  |  |
+---+--+--+--+--+--+--+--+--+--+--+--+--+--+--+--+
```

The 2 LI / E-octets

Highlighted octets / digits represent the MAC-content

Example of an RLC-AMD-PDU

Intentionally left blank

The STATUS- and Piggybacked STATUS-PDU (SUFI 1 – 4)

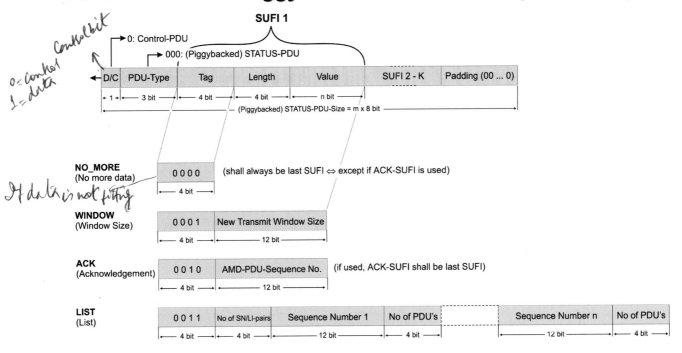

The STATUS- and Piggybacked STATUS-PDU (SUFI 1 – 4)

The STATUS-PDU can either be a "stand-alone" PDU to be transferred during an AM-data transfer between the two peers or the STATUS-PDU can be piggybacked within a regular AMD-PDU. In the latter case, the entire STATUS-PDU needs to fit into the available payload space of the AMD-PDU. Segmentation of STATUS-PDU's is not possible. As the figure illustrates, STATUS-PDU's may also require padding at the end.

Note:
- STATUS-PDU's are only applicable in case of AM.
- STATUS-PDU's are used to convey one or more SUFI's (Super Fields) to the peer which contain control information of different type.
- SUFI's are TLV-encoded (Tag Length Value). As the figure illustrates, the presence of Tag-, Length- and Value-field depends on the very SUFI.
- Every RLC-implementation needs to be able to interpret and react on all received SUFI-types. However, it is not required that every RLC-implementation can generate all SUFI-types (implementation option).

NO_MORE-SUFI
The NO_MORE-SUFI (⇔ No More Data) is used to indicate the end of the useful part of a STATUS-PDU (only if the ACK_SUFI is not used).

WINDOW-SUFI
Initially, RRC will configure the minimum and the maximum transmit window size and the receive window size which remains constant throughout the lifetime of an AM-connection. However, through the WINDOW-SUFI (⇔ Transmit Window Size) the receiver is able to readjust the transmit window of its peer within the given limits at any time.

ACK-SUFI
The ACK-SUFI (⇔ Acknowledgement) acknowledges positive reception for all AMD-PDU up to and including (AMD-PDU Sequence No – 1) that have not negatively been acknowledged by preceding SUFI's in the same STATUS_PDU. If used, the ACK-SUFI shall always be the last SUFI in a STATUS-PDU.

LIST-SUFI
The LIST-SUFI is used to efficiently request the retransmission of entire sections of consecutive erroneous AMD-PDU's from the peer. In that respect, each LIST-SUFI contains pairs of "Sequence Numbers" (12 bit) and "No of PDU's" fields (4 bit). In each pair, the "Sequence Number" points to the first erroneous AMD-PDU and "No of PDU's" indicates the number of consecutive AMD-PDU's which are also erroneous and need to be retransmitted.

[3GTS 25.322 (9.2.2.11)]

Example of a STATUS-PDU with SUFI's LIST and ACK

```
+----------+-----------------------------------+-----------------------------------------------+
|BITMASK   |ID Name                            |Comment or Value                               |
+----------+-----------------------------------+-----------------------------------------------+
|0010----  |2.1.1 MAC: C/T Field               |Logical Channel 3                              |
|          |2.1.2 MAC: Target Channel Type     |DCCH (Dedicated Control Channel)               |
|          |2.1.3 MAC: RLC Mode                |Acknowledge Mode                               |
|----0---  |2.1.4 RLC: Data/Control            |Control PDU                                    |
|-----000  |2.1.5 RLC: PDU Type                |STATUS                                         |
|2.1.6 RLC: List Super Field                                                                   |
|0011----  |2.1.6.1 RLC: SUFI Type             |List                                           |
|----0001  |2.1.6.2 RLC: Number of SN/L pairs  |1                                              |
|**b12***  |2.1.6.3 RLC: Sequence Number       |4                                              |
|----0000  |2.1.6.4 RLC: Numbers Of PDUs       |0 = PDU                                        |
|2.1.7 RLC: Acknowledgement Super Field                                                        |
|0010----  |2.1.7.1 RLC: SUFI Type             |Acknowledgement                                |
|**b12***  |2.1.7.2 RLC: Last Sequence Number  |6                                              |
|2.1.8 RLC: Padding                                                                            |
|**b100**  |2.1.8.1 RLC: Padding               |'000000000000000000000000000000000'B           |
|          |                                   | 000000000000000000000000000000000'B           |
|          |                                   | 000000000000000000000000000000000'B           |
+----+--+--+--+--+--+--+--+--+--+--+--+--+--+--+--+
|HEX |0 |1 |2 |3 |4 |5 |6 |7 |8 |9 |A |B |C |D |E |F |
+----+--+--+--+--+--+--+--+--+--+--+--+--+--+--+--+
|0   |d4|a4|01|20|31|00|40|20|06|00|00|00|00|00|00|00|
|10  |00|00|00|00|00|00|7f|91|  |  |  |  |  |  |  |  |
+----+--+--+--+--+--+--+--+--+--+--+--+--+--+--+--+
```

Please note that the numbering of sequence numbers is done separately for the different logical channels.

Highlighted octets / digits represent the MAC-content

Example of a STATUS-PDU with SUFI's LIST and ACK

Intentionally left blank

The STATUS- and Piggybacked STATUS-PDU (SUFI 5 – 8)

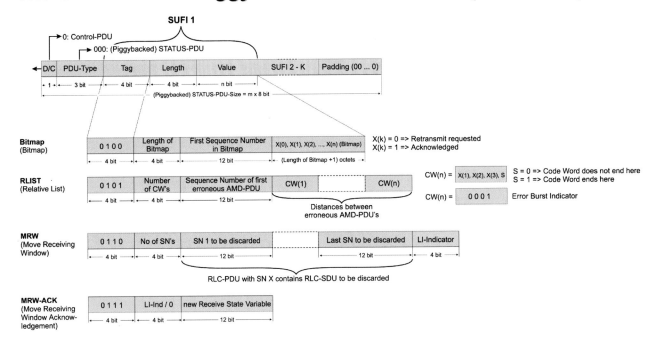

The STATUS- and Piggybacked STATUS-PDU (SUFI 5 – 8)

BITMAP-SUFI
The BITMAP-SUFI represents a standard ARQ-bitmap. The "Length of Bitmap"-field indicates the length of the bitmap in octets whereas the actual length of the bitmap is always ("Length of Bitmap" + 1). The following sequence number identifies the first AMD-PDU which is included in the bitmap (⇔ X(0)). The setting of the bits X(k) in the bitmap indicates whether an AMD-PDU is positively acknowledged or needs to be retransmitted.

RLIST-SUFI
The RLIST-SUFI is a quite sophisticated SUFI. It allows to efficiently point to scattered erroneous PDU's within the receive window or to identify entire sequences of erroneous PDU's. The CW's (Code Word) are used to identify the distances between erroneous PDU's or to identify the number of consecutive erroneous AMD-PDU's. To provide for higher code word value of '7', code words can be segmented over multiple CW's. A detailed example for the operation of the RLIST-SUFI can be found on the next page.

MRW-SUFI
The MRW-SUFI is used in case of RLC-SDU discard at the transmitter side to request the receiver to forward its receive window. Each included sequence number points to the very AMD-PDU which contains the final segment of the RLC-SDU to be discarded. The 4 bit long LI-Indicator informs the receiver which LI-field in the first in sequence AMD-PDU SN not to be discarded relates to the last RLC-SDU to be discarded.

MRW-ACK-SUFI
The MRW-ACK-SUFI is necessary to assure the re-synchronization of the transmit and receive windows after an RLC-SDU discard has happened. The new receive state variable points to the "oldest" SN that the receiver expects from the transmitter to be sent.

[3GTS 25.322 (9.2.2.11)]

Example: Operation of the RLIST-SUFI

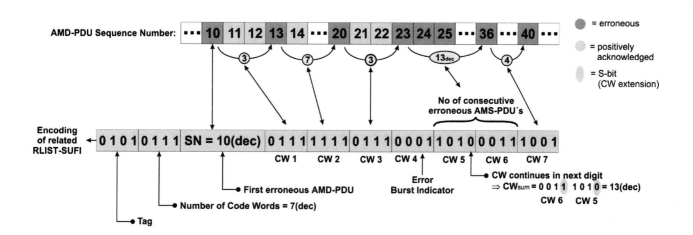

Example: Operation of the RLIST-SUFI

- **Explanation**
  - ⇒ The tag is set to '0101' = 5 to identify the RLIST-SUFI
  - ⇒ The number of 4 bit code words (CW) is '7'. Note that two next to last CW's 5 and 6 represent a combined code word but they are still counted as two separate CW's.
  - ⇒ The first erroneous AMD-PDU is the one with SN = 10_{dec}.
  - ⇒ The distance to the next erroneous AMD-PDU = 3 (⇔ SN = 13 needs to be retransmitted). Bit 0 (S-bit) in CW 1 indicates that this is the complete code word.
  - ⇒ The distance to the next erroneous AMD-PDU = 7 (⇔ SN = 20 needs to be retransmitted). Bit 0 (S-bit) in CW 2 indicates that this is the complete code word.
  - ⇒ The distance to the next erroneous AMD-PDU = 3 (⇔ SN = 23 needs to be retransmitted). Bit 0 (S-bit) in CW 3 indicates that this is the complete code word.
  - ⇒ CW 4 represents an "Error Burst Indicator". Consequently, the following code word shall be interpreted as the number of consecutive erroneous AMD-PDU's that follow AMD-PDU with SN = 23.
  - ⇒ The next two CW's are segmented. Bit 0 (S-bit) in CW 5 indicates that this code word continues in CW 6. Bit 0 (S-bit) in CW 6 indicates that this code word ends in CW 6. Altogether the meaning of CW 5 and CW 6 is therefore '001101'$_{bin}$ = 13_{dec}. Therefore, the AMD-PDU's with SN = 23 – 36 need to be retransmitted.
  - ⇒ Note that the next CW 7 is interpreted again as the distance towards the next erroneous AMD-PDU. Accordingly, an "Error Burst Indicator" is meaningful always only for the next, possibly segmented code word.

The RESET- and RESET-ACK-PDU

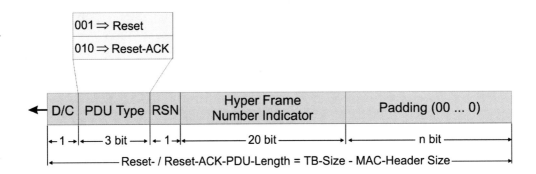

The RESET- and RESET-ACK-PDU

The RESET- and the RESET-ACK-PDU's are used to reset two RLC-entities which operate in AM. The reset-procedure can be triggered by 3 conditions:

- **A Maximum Number of Transmissions (⇔ MaxDAT) for an AMD-PDU has been performed and no Automatic RLC-SDU-Discard is Configured by RRC**
  The parameter MaxDAT is configured by RRC upon bearer setup. The range is MaxDAT = 1 ... 40_{dec}. Additionally, RRC may configure an automatic RLC-SDU discard function after MaxDAT transmissions of an AMD-PDU. In this case, the reset-procedure would not be invoked when an AMD-PDU has been transmitted MaxDAT-times. Note that MaxDAT does not relate to the number of retransmissions.

- **The Sender has Tried a Maximum Number of Times (⇔ MaxMRW) to Move the Receive Window in its Peer without Getting an Acknowledgement**
  Like MaxDAT, MaxWRW is configured by RRC upon bearer establishment. The range is MaxMRW = 1 ... 32_{dec}. If there is no reaction from the receiver after sending a STATUS-PDU with an embedded MRW-SUFI to the peer MaxMRW-times, then the reset-procedure is invoked by the sender.

- **A Received STATUS-PDU Indicated a Sequence Number out of the Current Receive Window**
  In this case, there is obviously a mismatch between sender and receiver. The reset-procedure is invoked to re-synchronize the two peers.

> **Note:**
> - After AM-bearer establishment, the RSN (Reset Sequence Number) shall be initialized to '0'. The expected RESET-ACK-PDU shall have the same RSN-value as the RESET-PDU. In case of retransmissions of the RESET-PDU, RSN shall not be incremented. A change of the RSN is only applicable, if there is more than one reset-procedure in the lifetime of an AM-bearer.
> - The HFNI (Hyper Frame Number Indicator) shall be set to the HFN that is currently used.

[3GTS 25.322 (11.4)]

Overflow Protection ⇔ RLC-SDU-Discard

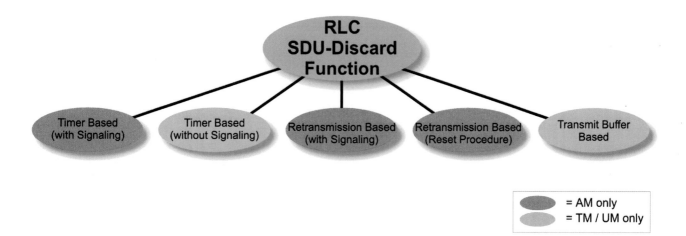

Overflow Protection ⇔ RLC-SDU-Discard

The RLC-SDU discard function is used by the RLC-sender to avoid an overflow situation. Several options for the RLC-SDU-discard function exist. For AM, the very procedure to be used is always configured by RRC. For TM and UM, the configuration of the RLC-SDU-discard function is optional. If nothing is configured, then RLC-SDU's shall be discarded, if the transmit buffer is full. The following options for the RLC-SDU-discard function are defined:

Timer Based RLC-SDU-Discard
Timer based RLC-SDU-discard uses the RLC-timer Timer_Discard to decide whether or not an RLC-SDU shall be discarded. Two options are defined.

With Signaling
This option is only applicable in RLC-AM. When an SDU is discarded, the MRW-SUFI is used to update the receiver.

Without Signaling
This option is only applicable in RLC-UM and RLC-TM. The sender will discard RLC-SDU for which the transmission doesn't succeed within Timer_Discard. In RLC-UM, the peer can be informed about the discard operation through incrementing the sequence number by '2' in the next UMD-PDU and indicating through the special length indicator that a new RLC-SDU starts with this UMD-PDU.

Retransmission Based RLC-SDU-Discard
Retransmission based RLC-SDU-discard is applicable only in RLC-AM. It uses the counter MaxDAT to determine when an RLC-SDU shall be discarded. Two options are defined:

With Signaling
An AMD-PDU which contains part or all of the affected RLC-SDU has been retransmitted (MaxDAT – 1) times. The entire RLC-SDU is discarded by the sender. The sender shall initiate the MRW-SUFI-procedure to update the receiver, if the sender supports the MRW-SUFI-procedure. Otherwise, option 2 (see below) applies:

Reset-Procedure
If an AMD-PDU which contains part or all of the affected RLC-SDU has been retransmitted (MaxDAT – 1) times, the entire RLC-SDU shall be discarded by the sender. In addition, the sender shall start the Reset-procedure.

Transmit Buffer Overflow Based RLC-SDU Discard
The default for the RLC-SDU-discard function in case of RLC-UM and RLC-TM. In case of RLC-TM, all received RLC-SDU's that have not been transmitted in one TTI, shall be discarded by the sender.

[3GTS 25.322 (9.7.3)]

Radio Resource Control (RRC)

- **Overview**

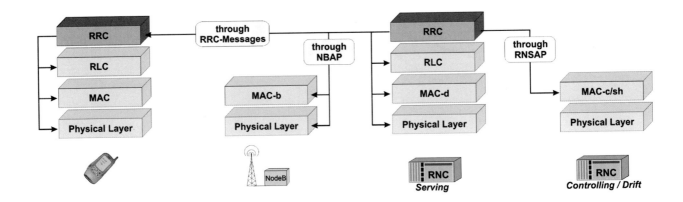

Radio Resource Control (RRC)

Overview
The figure provides an overview of RRC as the central and most important protocol within the UTRAN protocol stack.

- **Control of the UE Protocol Stack**
  The RRC-protocol within the SRNC controls most of the UE's protocol stack through the transfer of peer-to-peer RRC-signaling messages on BCCH and DCCH. The RRC-protocol within the CRNC communicates with the UE through CCCH. Note that the RRC-protocol also configures the physical layer within the UE.

- **Control of the NodeB Protocol Stack**
  The RRC-protocol within the CRNC controls the behavior of the NodeB (MAC-b / physical layer) by using the NBAP-protocol to transfer its requests to the NodeB.

- **Control of the SRNC / CRNC Protocol Stack**
  Obviously, every SRNC / CRNC controls its own protocol stack.

- **Partial Control of the DRNC Protocol Stack**
  In case of macrodiversity / soft handover situations, the SRNC partially controls the behavior of the DRNC and the affected NodeB's by using the RNSAP-protocol to transfer its requests to the DRNC.

[3GTS 25.331]

Tasks & Functions

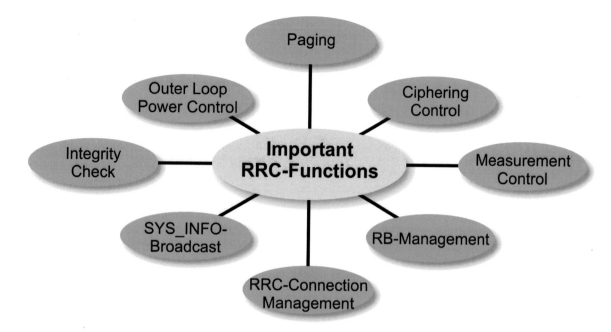

Tasks & Functions

- **Paging**
  RRC will relay the paging request from the core network to the UE. Depending on the RRC-state of the UE, either the PAG_TYPE1-message (⇔ RRC-idle, CELL_PCH, URA_PCH) or the PAG_TYPE2-message (⇔ CELL_DCH, CELL_FACH) shall be used.

- **Ciphering Control**
  RRC obtains ciphering data from the core network and the instruction whether ciphering shall be used. Unlike in GSM, the ciphering function is physically located in the UE and in the SRNC.

- **Measurement Control**
  The RRC-protocol in the RNC controls which measurements a UE shall perform. The measurement control is executed through system information (⇔ SIB 11, SIB 12) and through the MEAS_CTRL-message. Measurements are only applicable while the UE is in RRC-connected mode.

- **RB-Management (Radio Bearer)**
  RRC is entirely in charge for the management of the potentially multiple radio bearers of a single UE. Radio bearers need to be configured, possibly reconfigured and finally released.

- **RRC-Connection Management**
  Only upon request of the UE, but controlled by the RNC, an RRC-connection can be established between the UE and the SRNC. However, unless a radio link failure occurs, it is always the RNC that will release an RRC-connection when there is no more demand for it.

- **SYS_INFO Broadcast**
  The RRC-protocol in the CRNC controls which system information blocks with which content shall be broadcast by each connected cell. RRC uses the NBAP-protocol to convey the respective configuration towards the various NodeB's.

- **Integrity Check**
  The integrity check procedure is used to authenticate RRC-messages. During this procedure, the receiver confirms that an RRC-message has been sent by the correct RRC-peer.

- **Outer Loop Power Control**
  The RRC-protocol in the SRNC will check the received QE-values from the Iub-FP UL-DATA-frames and use the Iub-FP-message OUT_LOOP_PC to update the SIR-target within the NodeB. Consequentially, the NodeB will readjust the inner loop power control.

[3GTS 25.331]

Signaling Radio Bearers

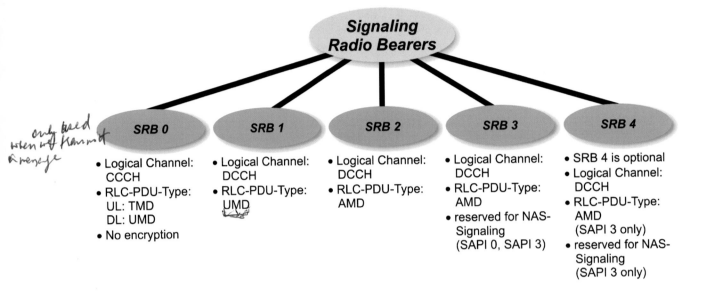

Signaling Radio Bearers

The figure illustrates the characteristics of:

Signaling Radio Bearer 0 (SRB 0)
SRB 0 can only be used on common control channels. In downlink direction, only UMD-PDU's may be used, in uplink direction only TMD-PDU's may be used. Information on SRB 0 will never be encrypted (neither in RLC nor in MAC).

Signaling Radio Bearer 1 (SRB 1)
SRB 1 is applicable only, if DCCH's are equipped. It uses the RLC-UM operation mode in uplink and downlink direction. Therefore, only UMD-PDU's are applied.

Signaling Radio Bearer 2 (SRB 2)
SRB 2 is applicable only, if DCCH's are equipped. Unlike SRB 1, SRB 2 uses the RLC-AM operation mode in uplink and downlink direction. Therefore, only AMD-PDU's can be applied.

Signaling Radio Bearer 3 (SRB 3)
SRB 3 is applicable only, if DCCH's are equipped. SRB 3 also uses the RLC-AM operation mode in uplink and downlink direction. SRB 3 is exclusively used to transfer piggybacked NAS-messages between UE and RNC. If SRB 4 is not equipped, then SRB 3 is used to transfer piggybacked NAS-messages related to SAPI 0 (⇔ CC, MM, RR) and SAPI 3 (⇔ SMS, SS). If SRB 4 is equipped (⇔ RNC-decision), then SRB 3 is exclusively used for the transfer of NAS-message that are related to SAPI 0

Signaling Radio Bearer 4 (SRB 4)
SRB 4 is optional. If it is equipped, it is exclusively used for the piggybacked transfer of NAS-messages that are related to SAPI 3 (⇔ SMS, SS).

> Note: All SRB's are exclusively used to transfer RRC-messages between the two RRC-peers.

[3GTS 25.331 (6.3)]

The Different RRC-States

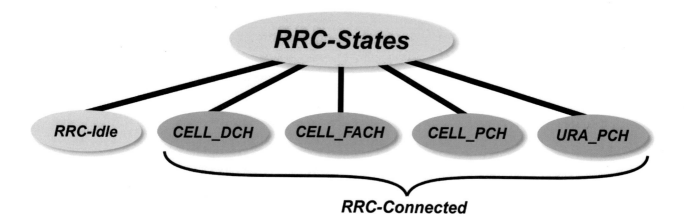

The Different RRC-States

Intentionally left blank

RRC-Idle Mode

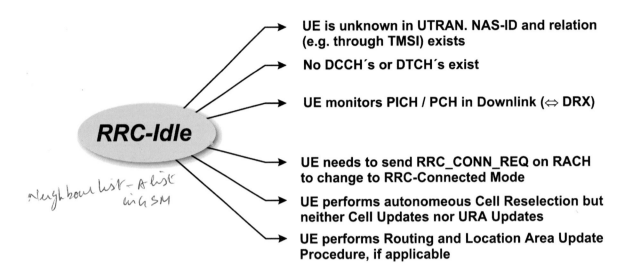

RRC-Idle Mode

UE is unknown in UTRAN
This statement relates in particular to the non-existence of RNTI's in the UTRAN for this UE. Still, the UE may be known in the circuit-switched and/or packet-switched core network, if it has previously registered (⇔ attachment). In this case, the respective core network entities will also have TMSI and/or P-TMSI allocated to the UE.

No DCCH's or DTCH's exist
Obviously, there can be no dedicated logical channels in RRC-idle mode.

UE monitors PICH / PCH in Downlink (⇔ DRX)
Hence, the core network can through UTRAN and the paging procedure request the UE to change to RRC-connected mode.

Change to RRC-Connected Mode requires Transmission of RRC_CONN_REQ
It is only the UE that can decide to trigger the RRC-connection establishment. This happens through the transmission of an RRC_CONN_REQ-message to the RNC. Obviously, the network can request the UE to trigger the RRC-connection establishment by sending a PAG_TYPE1-message on PCCH to the UE.

UE performs Autonomous Cell Reselection but neither Cell Updates nor URA Updates
While the UE is in RRC-idle, it will obviously reselect the serving cell based on radio interface criteria (⇔ CPICH-measurements) but the UE will not perform cell update or URA-update procedures.

UE Performs Routing and Location Area Update Procedures
If the new serving cell belongs to another routing or location area, the UE shall perform the respective MM-/GMM-procedures, if it is attached to the respective service.

[3GTS 25.304, 3GTS 25.331 (7.2.1)]

CELL_DCH-State

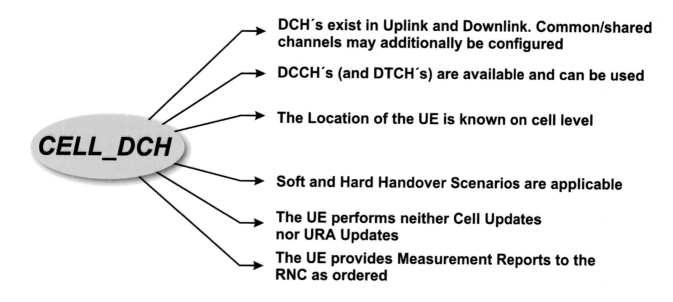

CELL_DCH-State

DCH's exist in Uplink and Downlink Direction
This is the most important characteristics of the CELL_DCH-state. Dedicated TrCH's are available in uplink and downlink direction and can be used for the exchange of DCCH- and, if configured, DTCH-logical channel information. Note that in CELL_DCH-state, common / shared TrCH's may be configured in addition, e.g. for the use by the DTCH's.

DCCH's are available and can be used; DTCH's may be available
Another important characteristics of the CELL_DCH state is that DCCH's are configured. Configuration relates to the mapping between DCCH X to DCH Y. DTCH's may also be configured in addition.
If UTRAN receives a paging message for the UE from the core network, it will use the PAG_TYPE2-message on DCCH (over DCH) to inform the UE. This is possible when the UE has radio access bearers established to one core network domain (e.g. circuit-switched) and UTRAN receives a paging from the other core network domain.

UTRAN knows the Location of the UE on Cell Level
The SRNC has allocated a U-RNTI to the UE and knows the location of the UE on cell level. The allocation of a C-RNTI is not required in CELL_DCH.

Handover Scenarios are Applicable
Only in CELL_DCH-state, soft and hard handover scenarios are applicable. Handover relates to the switching of the dedicated channels to another cell which may be a UTRAN-cell (FDD or TDD) or which may belong to another RAT (e.g. GSM).

UE performs no Cell Updates or URA Updates
In CELL_DCH-state the UE relies on the network to handover the UE to the best possible serving cell. Accordingly, neither cell updates nor URA-updates are applicable.

> **Exceptions:**
> 1. In case of a radio link failure or RLC-AM unrecoverable error in CELL_DCH-state, the UE shall initiate a cell update scenario with cause "radio link failure" and the network shall consecutively initiate the release of the RRC-connection by sending RRC_CONN_REL to the UE [3GTS 25.331 (8.3.1.2)].
> 2. In CELL_DCH-state, UTRAN may also initiate a cell change to a GERAN-cell by sending a CELL_CHAN_UTRAN-message to the UE, if the UE has only radio bearers established towards the packet-switched domain.

UE provides Measurement Reports to the RNC
Which measurements the UE performs depends on the setup information received from the RNC (⇔ SIB 11 and 12 / MEAS_CTRL-message).

[3GTS 25.331 (7.2.2.3, Annex B.3.1)]

CELL_FACH-State

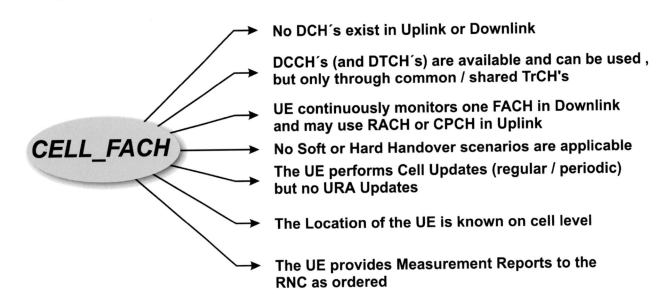

- No DCH's exist in Uplink or Downlink
- DCCH's (and DTCH's) are available and can be used, but only through common / shared TrCH's
- UE continuously monitors one FACH in Downlink and may use RACH or CPCH in Uplink
- No Soft or Hard Handover scenarios are applicable
- The UE performs Cell Updates (regular / periodic) but no URA Updates
- The Location of the UE is known on cell level
- The UE provides Measurement Reports to the RNC as ordered

CELL_FACH-State

No DCH's exist in Uplink or Downlink Direction
The most important difference between CELL_DCH- and CELL_FACH-state is that there are no dedicated TrCH's allocated to the UE in CELL_FACH-state.

DCCH's are available; DTCH's may be available
Like in CELL_DCH-state, DCCH exist in CELL_FACH-state and DTCH's may exist. However, since no DCH's are available, these logical channels are linked to common / shared transport channels.

UE continuously monitors one FACH in Downlink
Since there are no dedicated TrCH's available in CELL_FACH-state, the UE shall continuously monitor one FACH-TrCH to receive information on DCCH's and, if configured, on DTCH's which are destined for that UE. The UE can be addressed through the U-RNTI or the C-RNTI.
If the UE needs to transmit information on DCCH or DTCH in uplink direction, it shall use the RACH- or CPCH-TrCH. Accordingly, the UE needs to be assigned a C-RNTI in CELL_FACH-state.
If UTRAN receives a paging message for the UE from the core network, it will use the PAG_TYPE2-message on DCCH (over FACH) to inform the UE.

No Soft or Hard Handover Scenarios are applicable
There are no handover scenarios possible, since the UE occupies no dedicated transport channels. It is the responsibility of the UE to select the most suitable cell and to perform the related cell update scenario. Still, through the CELL_CHAN_UTRAN-message, also the UTRAN is able to direct the UE to a GSM-cell.

UE performs Cell Updates but no URA Updates
In CELL_FACH-state, the UE will autonomously reselect the serving cell and perform a cell update scenario consecutively. Cell updates will also performed periodically (if configured) upon timer T305 expiry.

UTRAN knows the Location of the UE on Cell Level
Even in CELL_FACH-state, the SRNC (and possible another CRNC) knows the location of the UE on cell level, because the UE performs cell update scenarios.

UE provides Measurement Reports to the RNC
Which measurements the UE performs depends on the setup information received from the RNC (⇔ SIB 11 and 12 / MEAS_CTRL-message).

[3GTS 25.331 (7.2.2.2, Annex B.3.2)]

CELL_PCH-State

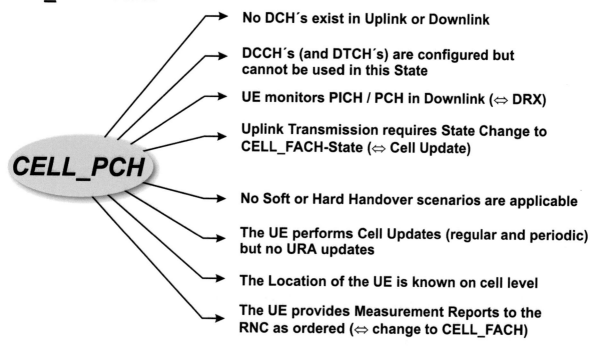

CELL_PCH-State

No DCH's exist in Uplink and Downlink Direction
Like in CELL_FACH-state there are no dedicated resources allocated to the UE in CELL_PCH-state.

DCCH's (and DTCH's) are configured but cannot be used in this State
The most important difference between CELL_FACH-state and CELL_PCH-state is that in CELL_PCH-state there are no DCCH's nor DTCH's available.

UE monitors PICH / PCH in Downlink (⇔ DRX)
If the network needs to transmit information to the UE it needs to page it through a PAG_TYPE1-message (in the cell and on PCCH / PCH) prior to sending any useful information. Accordingly, the UE shall monitor the PICH to detect paging on PCH which are destined for that UE. The reception of a PAG_TYPE1-message will trigger a cell update scenario (cause: paging response) as paging response.

Uplink Transmission requires State Change to CELL_FACH (⇔ Cell Update)
If the UE needs to transmit uplink data or signaling information, it needs to perform a cell update scenario (cause: uplink data transmission) prior to any data transfer.

No Soft or Hard Handover Scenarios are applicable
There are no handover scenarios possible, since the UE occupies no dedicated transport channels. If the UTRAN wants to move the UE to another cell, it prior needs to page it and move it to CELL_FACH-state.

UE performs Cell Updates but no URA Updates
In CELL_PCH-state, the UE will autonomously reselect the serving cell and perform a cell update scenario consecutively. Cell updates will also performed periodically (if configured) upon timer T305 expiry. However, no URA-Updates are applicable.

UTRAN knows the Location of the UE on Cell Level
Even in CELL_PCH-state, the SRNC (and possible another CRNC) knows the location of the UE on cell level, because the UE performs cell update scenarios.

UE provides Measurement Reports to the RNC
Which measurements the UE performs depends on the setup information received from the RNC (⇔ SIB 11 and 12 / MEAS_CTRL-message). The transmission of a MEAS_REP-message requires the previous state change to CELL_FACH and therefore a cell update scenario.

[3GTS 25.331 (7.2.2.1, Annex B.3.3)]

URA_PCH-State

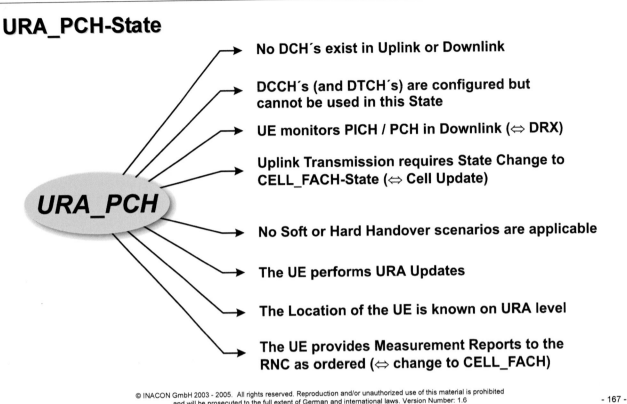

URA_PCH-State

No DCH's exist in Uplink and Downlink Direction
Like in CELL_PCH-state there are no dedicated resources allocated to the UE in URA_PCH-state.

DCCH's (and DTCH's) are configured but cannot be used in this State
Like in CELL_PCH-state the DCCH's and possibly DTCH's are configured but cannot be used in URA_PCH-state.

UE monitors PICH / PCH in Downlink (⇔ DRX)
If the network needs to transmit information to the UE it needs to page it through a PAG_TYPE1-message (in the URA and on PCCH / PCH) prior to sending any useful information. Accordingly, the UE shall monitor the PICH to detect paging on PCH which are destined for that UE. The reception of a PAG_TYPE1-message will trigger a cell update scenario (cause: paging response) as paging response.

Uplink Transmission requires State Change to CELL_FACH (⇔ Cell Update)
If the UE needs to transmit uplink data or signaling information, it needs to perform a cell update scenario (cause: uplink data transmission) prior to any data transfer.

No Handover Scenarios are Applicable
There are no handover scenarios possible, since the UE occupies no dedicated transport channels.

UE performs URA Updates
Most likely the network will move a UE from CELL_PCH- or CELL_FACH-state to URA_PCH-state after some inactivity time (network dependent) to avoid continuous cell updates of that UE. However, although the UE shall not perform cell updates due to cell reselection or periodic updating in URA_PCH-state, it shall perform URA-updates upon cell reselection to a cell which belongs to another URA.

UTRAN knows the Location of the UE on URA Level
Accordingly, the UTRAN will know the UE's location only on URA-level.

UE provides Measurement Reports to the RNC
Which measurements the UE performs depends on the setup information received from the RNC (⇔ SIB 11 and 12 / MEAS_CTRL-message).

[3GTS 25.331 (7.2.2.1, Annex B.3.4)]

RRC State Transitions and Transitions to/from GSM

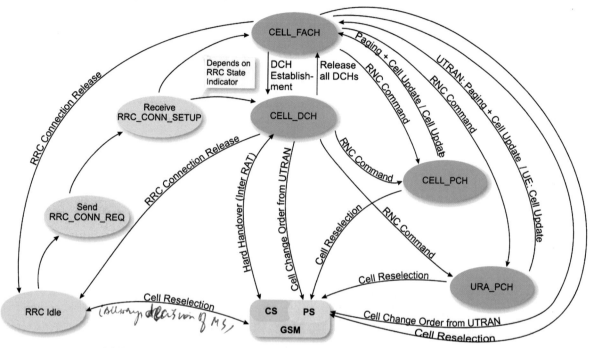

RAB → Radio Access Bearer.

RRC State Transitions and Transitions to/from GSM

The figure illustrates the possible transitions RRC-Idle ⇔ RRC-Connected ⇔ GSM

- **Some Remarks**
  ⇒ When the UE is in RRC-idle mode, it may need to transmit uplink signaling or data which requires the establishment of an RRC-connection which consecutively triggers the transmission of an RRC_CONN_REQ-message.
  ⇒ If there is no more suitable UTRAN-cell while the UE is in RRC-idle mode, the UE may at any time perform a cell reselection to a GSM-cell, if the UE is capable to support GSM (multi-mode terminal). Whether or not a cell reselection from RRC-idle mode to GSM requires a routing and/or location area updating scenario, depends on whether or not the new GSM serving cell belongs to another routing / location area than the UTRAN-cell.
  ⇒ The RRC_CONN_SETUP will always contain the IE "RRC-State-Indicator" which tells the UE to move into either CELL_DCH or CELL_FACH-state.
  ⇒ Upon release of the last DCH-TrCH, the UE will autonomously move into CELL_FACH-state, unless the SRNC has explicitly ordered the UE to move into CELL_PCH or URA_PCH-state or to release the RRC-connection, altogether.
  ⇒ While in CELL_FACH-state, the UE may autonomously select a GSM-cell as serving cell or it may receive a CELL_CHAN_UTRAN-message from the SRNC to do so. In this case, the UE definitely needs to perform a GPRS cell update scenario and possibly a routing and/or a location area updating scenario. The same applies in CELL_DCH-state, if the UE receives a CELL_CHAN_UTRAN-message.
  ⇒ In CELL_DCH-state, the CELL_CHAN_UTRAN-message is only applicable, if the UE has only RAB's allocated towards the packet-switched core network domain. In all other cases, the HO_UTRAN_GSM-message is required to handover the UE to a GSM-cell.
  ⇒ While in CELL_PCH-state and URA_PCH-state, the UE is unable to transmit any data or signaling information. If e.g. a measurement report needs to be transmitted, the UE needs to perform a cell update scenario prior to sending this measurement report.
  ⇒ While the UE is in CELL_PCH- or URA_PCH-state, a cell reselection to a GSM serving cell requires a GPRS cell update scenario and possibly also a routing and/or location updating scenario.

Paging:
- While the UE is in RRC-idle mode, CELL_PCH- and URA_PCH-state, the UTRAN will use the PAG_TYPE1-message and the PCCH / PCH to page the UE, if requested by either core network domain.
- While the UE is in CELL_DCH-state or CELL_FACH-state, the UTRAN will use the PAG_TYPE2-message and the DCCH / appl. TrCH to page the UE, if requested by either core network domain.

[3GTS 25.331 (Annex B)]

Encoding of RRC-Messages

- ASN.1: Basic Encoding Rules (BER)

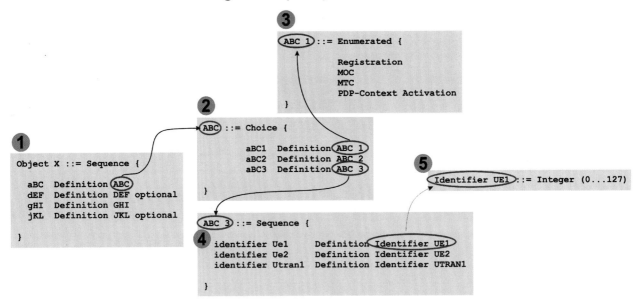

Encoding of RRC-Messages

ASN.1: Basic Encoding Rules (BER)

Nested Information Elements
The most important hint for the interpretation of ASN.1 code is illustrated in the figure: ASN.1 uses a hierarchical description form. One information element is nested into another which again may be nested into another information element which …

ASN.1 Types
The figure illustrates only a fraction of the different description types that ASN.1 is using. For a complete listing please refer to the respective standards (see references underneath).

The Sequence Type
The sequence type describes which information elements together form an information element. In the figure, Object X (⇔ point (1)) consists of the IE's ABC, DEF, GHI and JKL. Note that DEF and JKL are indicated as being optional. The figure also illustrates that ABC3 (⇔ point (4)) is again a sequence of different identifiers.

The Choice Type
The choice type is a "one out of many" descriptor. In the example (⇔ point (2)), the IE ABC is a choice of either ABC1 or ABC2 or ABC3. Only one can be present as ABC in Object X.

The Enumerated Type
The enumerated type provides a simple identifier for something. In the figure (⇔ point (3)), ABC1 is either registration = 1 or MOC = 2 or MTC = 3 or PDP-Context-Activation = 4. Only one these values will be included in the message.

Value Assignment
In the figure (⇔ point (5)), IdentifierUE1 is assigned an integer value with a range between '0' and '127'$_{dec}$.

> Note: ASN.1 encoding is used in various GSM- and UMTS-protocols like MAP, NBAP, RANAP and RNSAP.

[ITU-T X.680, X.681]

TLV-Encoding

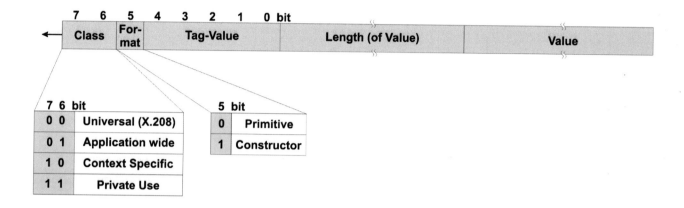

TLV-Encoding

Another important aspect of the ASN.1-description is the TLV-encoding (Tag, Length, Value). Each information element is encoded by the ASN.1-compiler defining a tag-value and a length-value for it. The figure illustrates the simplest possibility when tag- and length-fields are only 8 bit long. Later in this section, we will present how these fields can be extend to match the need for higher tag-values and longer lengths.

The Tag-Field
The tag-field is again split into three parts:

- **The Class Field**
The class field is used to distinguish IE's depending on their validity in different environments. For instance, the tag-value for an IE which is used in a private network is ambiguous. The same private tag-value may be used in other private networks, too.

Class	Meaning	Description
00	Universal	Universal data types as specified by the ITU-T in the recommenda-tion X.208 (e.g. Integer, Boolean, Sequence, Enumerated, ...)
01	Application Wide	Only valid within a specific set of ITU-T recommendations (e.g. TCAP-specific IE's (⇔ Q.771, Q.772, Q.773))
10	Context Specific	Only valid within a specific ITU-T application (e.g. GSM-MAP)
11	Private Use	Network or corporate specific assignments that will never be considered by the ITU-T.

- **The Format Field**
It was mentioned earlier: ASN.1 provides for nested information elements. The "format field" is used to distinguish whether an IE is not further divided into other IE's (⇔ primitive / format = 0) or whether an IE consists of further IE's that are nested into it (⇔ constructor / format = 1).

- **The Tag-Value**
Tag values for the universal class tags have been allocated by the ITU-T in recommendation X.208. Other recommendations contain the defined tag values for application wide and context specific tags.

The Length Field
The length field indicates the length of the value field in octets.

[ITU-T X.680, X.681]

Example: Encoding of IMSI

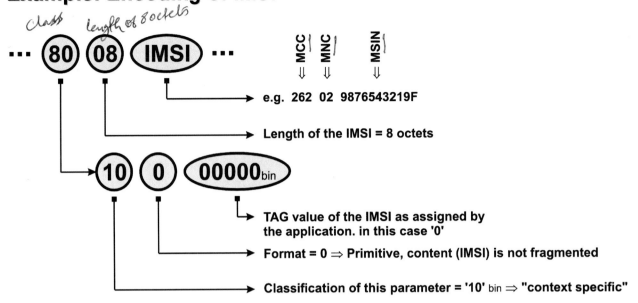

Example: Encoding of IMSI

⇒ In the example, the IE "IMSI" shall be ASN.1-encoded and represented in a message.
⇒ MCC, MNC and MSIN shall not be considered as separate IE's, hence the IMSI becomes a primitive (format = 0_{bin})
⇒ In the example, the IMSI is MCC = 262, MNC = 02, MSIN = 9876543219F
⇒ The length of the IMSI is therefore 8 octets (length field = 8).
⇒ In the example, the IMSI is considered a context specific IE (⇔ GSM-MAP). Therefore, the class = 10_{bin}.

Extension of Tag- and Length-Fields

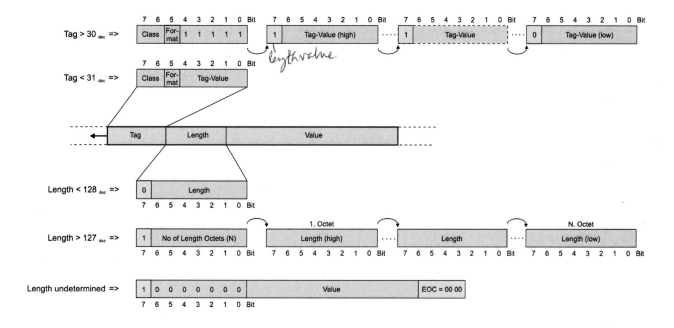

Extension of Tag- and Length-Fields

As the figure illustrates it is obviously possible to extend the Tag- and Length-Fields beyond a length of 1 octet each.

Extension of the Tag-Value (Tag-Values > 30dec)
If tag-values > 30dec need to be allocated, an extension mechanism for the tag-field has been defined. As illustrated, the first octet of the tag-field will still contain the class- and format information. The tag-value '11111'$_{bin}$ in this first octet identifies the tag extension but is not part of the actual tag-value. This tag value starts with the next octet or rather with the least significant 7 bit of the next octet. If the tag-value continues in the next octet, the most significant bit of this octet and the following octets shall be set to '1'. If an octet contains the final (least significant) part of the tag-value, then the most significant bit of this octet shall be set to '0'.

Extension of the Length Value (Length Value > 127dec)
The extension mechanism for the length field works in a similar fashion. The most significant bit 7 of the first length octet indicates whether the length field is spread over at least 2 octets. In such a case, the least significant 7 bit of this first octet indicate the number of octets that follow and that contain the actual length information of the value-field. In other words: The length field can have an overall length of 1 octet up to 128 octets.

Undetermined Length
For some IE's, the length cannot be determined during compilation. To cope with such situations, the ITU-T has defined a special length indicator for undetermined length. Bit 7 of the length octet shall be set to '1' and bit 0 – 6 will be '000 000'. Then the data part follows. The end of the data part will be indicated through two EOC-octets = 00 00 (End Of Code).

[ITU-T X.680, X.681]

ASN.1: Packed Encoding Rules (PER) / Unaligned

- **Erase Tag and Length Fields**

ASN.1: Packed Encoding Rules (PER) / Unaligned

Note:
- The Packed Encoding Rules are used to minimize and compress the output production of the basic encoding rules into a more efficient presentation format.
- PER exist in two versions: Aligned and unaligned format. For the RRC-protocol, the unaligned version is used. NBAP, RANAP and RNSAP use the aligned format.
- **Unaligned Format:** The bit-field output of the different IE's is concatenated without any padding.
- **Aligned Format:** The bit-field output of an IE is padded with '0'-bits to an octet boundary before concatenation.
- The following topics only provide the most important rules of PER that are required to analyze and interpret binary and mnemonic RRC-, NBAP-, RANAP- and RNSAP-recording files. For more details please refer to X.691.

Erase Tag and Length Fields
If an IE is mandatory, why spend at least 1 octet each for tag and length field. When PER are applied, the ASN.1-encoded IE's are stripped from their tag- and length fields.

[ITU-T X.691]

Minimize the Value Field

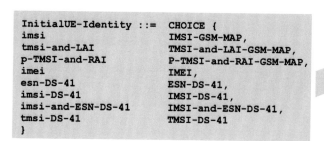
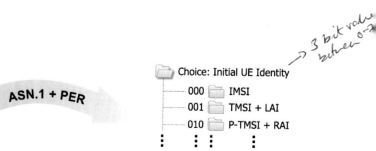

ASN.1 + PER

→ 3 bit value between 0-7

TMSI → Terminal (MS)
IMSI → Sim card.

Minimize the Value Field

If there are only 8 different values defined for a given IE, why waste 8 bit as the minimum length of the value field? PER will evaluate how many values are defined for any given IE. If possible, the reserved number of bits will be reduced to the absolute minimum.

- **Example**
  The graphics page illustrates the unaligned PER-encoding of the IE "Initial UE Identity". Since there are only 8 different possibilities defined, 3 bit are sufficient to identify which of the different UE-ID's is used.
  ⇒ If IMSI is used, the encoding will be {000 IMSI}
  ⇒ If TMSI + LAI is used, the encoding will be {001 TMSI LAI}
  ⇒ and so on.

[ITU-T X.691, 3GTS 25.331]

Handling of Optional IE's

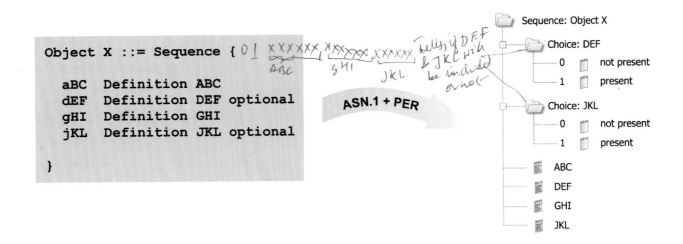

Handling of Optional IE's

If an IE consists of the sequence type and contains one or more optional IE's and IE's with default values, then a bitmap at the beginning of the packed ASN.1-output for the sequence type IE shall indicate whether or not an optional IE is present.
In our example, the bitmap consists of only 2 bit which indicate whether or not the optional parameters DEF and/or JKL are present.

[ITU-T X.680, X.681, X691]

Handling of IE's with Variable Length

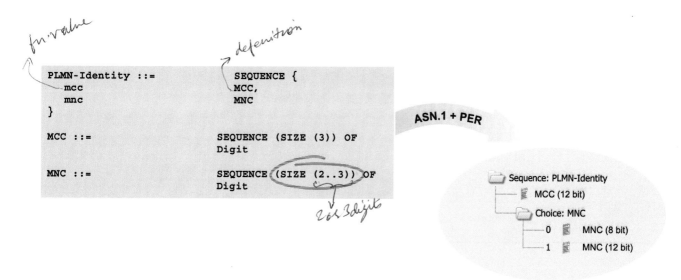

Handling of IE's with Variable Length

If an information element comes with a variable length, PER need to provide for a possibility to identify the current length of this IE. A simple example is illustrated on the graphics page:
Encoding of the MNC (Mobile Network Code)
The MNC may be 2 or 3 digits long. Each digit consists of 4 bit.

[ITU-T X.691]

Encoding Example: The RRC_CONN_REQ-Message

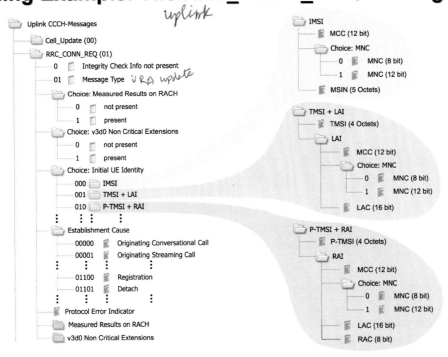

Encoding Example: The RRC_CONN_REQ-Message

Intentionally left blank

UMTS – Signaling & Protocol Analysis

Practical Exercise:

- Decode an RRC_CONN_REQ-Message   *(Always transmitted on CCCH as logical ch.)*
  RACH   *(ACK & UnAck, both allowed)*

9 octets is message

```
+----+--+--+--+--+--+--+--+--+--+--+--+--+--+--+--+--+
|HEX | 0| 1| 2| 3| 4| 5| 6| 7| 8| 9| A| B| C| D| E| F|
+----+--+--+--+--+--+--+--+--+--+--+--+--+--+--+--+--+
|  0 |21|3e|00|1c|6b|26|23|b0|01|eb|00|00|00|00|00|00|
| 10 |00|00|00|00|00|  |  |  |  |  |  |  |  |  |  |  |
+----+--+--+--+--+--+--+--+--+--+--+--+--+--+--+--+--+
```

Initial UE identity → IMSI
→ P-TMSI + RAI

Answers:

Initial UE Identity Type = _____ / in hex:

Message type → no ext. included
element not included
21 — 0010 0001 → TMSI + LAI (Identifier)
3e = 0011 1110 } TMSI
0000 0000
0001 1100
0101 1011
0010 0101 → Mobile Network code in 8 bit
0010 0011
1011 0000 } LAC
0000 0001 → Registration
1110 1011
Padding → 0000 0000

Protocol Error Indicator

Establishment Cause = _____

RRC-Message Types

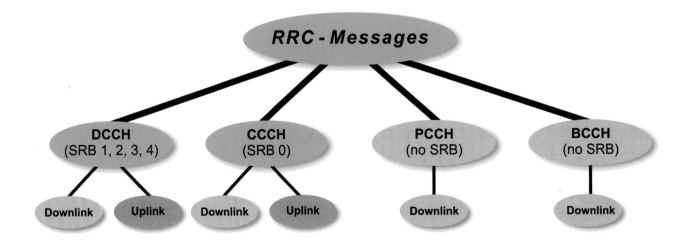

RRC-Message Types

As the figure illustrates, RRC-message are separated into different groups, depending on the logical channel type

- **RRC-Messages on DCCH**
  On DCCH, another separation is given by the direction: There are uplink and downlink RRC-messages on DCCH. Signaling radio bearer 1, 2, 3 and 4 are applicable.
- **RRC-Messages on CCCH**
  On CCCH, another separation is given by the direction: There are uplink and downlink RRC-messages on CCCH. Only signaling radio bearer 0 is applicable.
- **RRC-Messages on PCCH**
  On PCCH, there are only downlink RRC-messages. When the PCCH is used, there is no signaling radio bearer.
- **RRC-Messages on BCCH**
  On BCCH, there are only downlink RRC-messages. When the BCCH is used, there is no signaling radio bearer.

Downlink DCCH-Messages

Message Type 5 bit / [bin]	RLC-Mode & Signal. Radio Bearer (SRB)	Messages on Downlink DCCH Name (ASN.1)	Abbreviation
00000	UM / AM (SRB 1 / 2)	activeSetUpdate	ACT_SET_UPD
00001	AM (SRB 2)	assistanceDataDelivery	ASSIST_DATA_DEL
00010	AM (SRB 2)	cellChangeOrderFromUTRAN	CELL_CHAN_UTRAN
00011	UM (SRB 1) *)	cellUpdateConfirm	CELL_UPD_CNF *)
00100	AM (SRB 2)	counterCheck	COUNT_CHECK
00101	AM (SRB 3 / 4)	downlinkDirectTransfer	DL_DIR_TRANS
00110	AM (SRB 2)	handoverFromUTRANCommand-GSM	HO_UTRAN_GSM
00111	AM (SRB 2)	handoverFromUTRANCommand-CDMA2000	HO_UTRAN_CDMA2000

*) Message is also defined for downlink CCCH

Downlink DCCH-Messages

ACT_SET_UPD
The ACT_SET_UPD-message is sent to the UE during soft and softer handover to add, replace or delete radio links within the active link set of the UE. Note that one ACT_SET_UPD-message may simultaneously add and delete links from the active link set.

ASSIST_DATA_DEL
The ASSIST_DATA_DEL-message is sent to the UE to provide GPS- or OTDOA-related positioning information to the UE.

CELL_CHAN_UTRAN
The CELL_CHAN_UTRAN-message is sent to the UE in CELL_DCH-state or CELL_FACH-state to order a cell reselection of the UE to another radio access technology, most likely to a GSM-PLMN. This message may only be used if there only radio access bearers towards the packet-switched domain or if there are no radio access bearers allocated.

CELL_UPD_CNF
The SRNC conveys this message on DCCH to the UE to provide for ciphering. The intention of the message is to confirm a cell update scenario and possibly to change / add the physical and/or transport channel information and to allocate new RNTI's (C-RNTI, U-RNTI).

COUNT_CHECK
The SRNC may use the COUNT_CHECK-message to retrieve the data volume counters from the UE for sent and received data on (data) radio bearers. The SRNC will do so to confirm whether or not the reported data volumes match the data volumes measured in the network (⇔ man in the middle attack). This procedure is only applicable for radio bearers in RLC-UM and RLC-AM-operation mode.

DL_DIR_TRANS
The DL_DIR_TRANS-message is used by the SRNC to transmit NAS-messages to the UE.

HO_UTRAN_GSM
The HO_UTRAN_GSM-message is used by the SRNC to order a UE to perform a hard handover to a GSM-cell.

HO_UTRAN_CDMA_2000
The HO_UTRAN_CDMA_2000-message is used by the SRNC to order a UE to perform a hard handover to a cdma2000-cell.

[3GTS 25.331]

Downlink DCCH-Messages

Message Type 5 bit / [bin]	RLC-Mode & Signal. Radio Bearer (SRB)	Messages on Downlink DCCH Name (ASN.1)	Abbreviation
01000	AM (SRB 2)	measurementControl	MEAS_CTRL
01001	AM (SRB 2)	pagingType2	PAG_TYPE2
01010	UM / AM (SRB 1 / 2)	physicalChannelReconfiguration	PHYS_CHAN_RECONF
01011	UM (SRB 1)	physicalSharedChannelAllocation	PHYS_SHCH_ALL
01100	UM / AM (SRB 1 / 2)	radioBearerReconfiguration	RB_RECONF
01101	UM / AM (SRB 1 / 2)	radioBearerRelease	RB_REL
01110	UM / AM (SRB 1 / 2)	radioBearerSetup	RB_SETUP
01111	UM (SRB 1) *)	rrcConnectionRelease	RRC_CONN_REL *)

*) Message is also defined for downlink CCCH

Downlink DCCH-Messages

MEAS_CTRL
The MEAS_CTRL-message is used by the SRNC to initiate, modify or stop specific measurements in the UE. Note that the UE receives a standard measurement setup already through SIB 11 / 12. This standard measurement setup is only further specified by the MEAS_CTRL-message.

PAG_TYPE2
The PAG_TYPE2-message is only applicable, if the SRNC receives a paging request from either core network domain for a UE which is either in CELL_DCH-state or in CELL_FACH-state. The UE shall respond to a PAG_TYPE2-message with a NAS paging response message (e.g. (PAG_RSP)) over the existing RRC-connection and addressed to the respective core network domain.

PHYS_CHAN_RECONF
In general, the PHYS_CHAN_RECONF-message is used by the SRNC to change the physical channel configuration of a UE (e.g. used scrambling code). In addition, the PHYS_CHAN_RECONF-message is the standard message to initiate a hard handover FDD ⇒ FDD or FDD ⇒ TDD.

PHYS_SHCH_ALL
(TDD only)

RB_RECONF
The RB_RECONF-message is used by the SRNC mainly to reconfigure parameters of the possibly different radio bearers of a UE. It can also be used to initiate a hard handover procedure FDD ⇒ FDD or FDD ⇒ TDD.

RB_REL
The RB_REL-message is used by the SRNC to release the active radio bearers of a UE. This does not affect the RRC-connectivity.

RB_SETUP
The RB_SETUP-message is used by the SRNC to establish radio bearers for user plane data transfer for one UE.

RRC_CONN_REL
The RRC_CONN_REL-message is used by the SRNC to release the RRC-connection of a UE. This message is sent on DCCH, if when the UE is in CELL_DCH- or CELL_FACH-state and the DCCH is available. If sent on DCCH, the message shall be answered by the UE using the RRC_CONN_REL_COM-message.

Downlink DCCH-Messages

Message Type 5 bit / [bin]	RLC-Mode & Signal. Radio Bearer (SRB)	Messages on Downlink DCCH Name (ASN.1)	Abbreviation
10000	AM (SRB 2)	securityModeCommand	SEC_MODE_CMD
10001	AM (SRB 2)	signallingConnectionRelease	SIG_CONN_REL
10010	UM / AM (SRB 1 / 2)	transportChannelReconfiguration	TrCH_RECONF
10011	UM / AM (SRB 1 / 2)	transportFormatCombinationControl	TFC_CONTROL
10100	UM / AM (SRB 1 / 2)	ueCapabilityEnquiry	UE_CAP_ENQ
10101	UM / AM (SRB 1 / 2)	ueCapabilityInformationConfirm	UE_CAP_INF_CNF
10110	UM / AM (SRB 1 / 2)	uplinkPhysicalChannelControl	UL_PHYS_CHAN_CTRL
10111	UM (SRB 1) *)	uraUpdateConfirm	URA_UPD_CNF *)

*) Message is also defined for downlink CCCH

Downlink DCCH-Messages

SEC_MODE_CMD
The SEC_MODE_CMD-message is used to start or adjust the integrity protection procedure for all SRB's (1, 2, ...) and to start or adjust the ciphering procedure for the radio bearers of one core network domain and for all SRB's (1, 2, ...).

SIG_CONN_REL
The SIG_CONN_REL-message is used to inform the UE that one of its ongoing signaling connections to either core network domain has been released. Note that this does not relate to the release of the RRC connection. The UE can also trigger the signaling connection release by sending a SIG_CONN_REL_IND-message to the SRNC on uplink DCCH.

TrCH_RECONF
The TrCH_RECONF-message is used by the SRNC mainly to reconfigure parameters of the possibly different TrCH's of a UE. It can also be used to initiate a hard handover procedure FDD ⇒ FDD or FDD ⇒ TDD.

TFC_CONTROL
The TFC_CONTROL-message is used by the SRNC to expand or limit the allowed uplink TFC's within the configured TFCS.

UE_CAP_ENQ
The UE_CAP_ENQ-message is used by the SRNC to request the UE to transmit its radio capabilities related to any radio access technology to the SRNC.

UE_CAP_INF_CNF
The UE_CAP_INF_CNF-message is used by the SRNC to confirm the reception of the UE_CAP_INFO-message.

UL_PHYS_CHAN_CTRL
(TDD only)

URA_UPD_CNF
The URA_UPD_CNF-message is sent by the SRNC to the UE to confirm a successful URA-update scenario.

Downlink DCCH-Messages

Message Type 5 bit / [bin]	RLC-Mode & Signal. Radio Bearer (SRB)	Messages on Downlink DCCH Name (ASN.1)	Abbreviation
11000	UM / AM (SRB 1 / 2)	utranMobilityInformation	UTRAN_MOB_INFO
11001	-/-	spare7	-/-
11010	-/-	spare6	-/-
11011	-/-	spare5	-/-
11100	-/-	Spare4	-/-
11101	-/-	spare3	-/-
11110	-/-	spare2	-/-
11111	-/-	spare1	-/-

Downlink DCCH-Messages

UTRAN_MOB_INFO
The UTRAN_MOB_INFO-message is used by the SRNC to allocate new C-RNTI or U-RNTI to a UE while the UE is in RRC-connected mode. Other information which may be included are e.g. new timer / counter values.

Note: Spare-values are reserved for future use.

UMTS – Signaling & Protocol Analysis

Uplink DCCH-Messages

Message Type 5 bit / [bin]	RLC-Mode & Signal. Radio Bearer (SRB)	Messages on Uplink DCCH Name (ASN.1)	Abbreviation
00000	AM (SRB 2)	activeSetUpdateComplete	ACT_SET_UPD_COM
00001	AM (SRB 2)	activeSetUpdateFailure	ACT_SET_UPD_FAIL
00010	AM (SRB 2)	cellChangeOrderFromUTRANFailure	CELL_CHAN_UTRAN_FAIL
00011	AM (SRB 2)	counterCheckResponse	COUNT_CHECK_RSP
00100	AM (SRB 2)	handoverToUTRANComplete	HO_UTRAN_COM
00101	AM (SRB 3 / 4)	initialDirectTransfer	INIT_DIR_TRANS
00110	AM (SRB 2)	handoverFromUTRANFailure	HO_UTRAN_FAIL
00111	AM (SRB 2)	measurementControlFailure	MEAS_CTRL_FAIL

UMTS – Signaling & Protocol Analysis

Uplink DCCH-Messages

ACT_SET_UPD_COM
The ACT_SET_UPD_COM-message is sent by the UE to confirm the changes in the active link set previously received in an ACT_SET_UPD-message from the SRNC.

ACT_SET_UPD_FAIL
The UE shall send an ACT_SET_UPD_FAIL-message to the SRNC, if it cannot properly react on a received ACT_SET_UPD-message. Possible reasons are for instance unsupported link configuration or incompatible spreading factors. After sending ACT_SET_UPD_FAIL, the UE shall remain on the previous link set and wait for further instructions.

CELL_CHAN_UTRAN_FAIL
The UE will send a CELL_CHAN_UTRAN_FAIL-message to the SRNC, if a network initiated cell update procedure to GPRS failed. One possible reason is the expiry of T309 before access to GPRS is successful.

COUNT_CHECK_RSP
If requested by the SRNC through a COUNT_CHECK-message, the UE shall convey a COUNT_CHECK_RSP-message to the SRNC, indicating the traffic transfer volume for all established radio bearers in RLC-UM and RLC-AM mode.

HO_UTRAN_COM
The HO_UTRAN_COM-message is sent during a hard handover procedure from GSM to UTRAN to confirm the successful radio bearer establishment in UTRAN.

INIT_DIR_TRANS
The INIT_DIR_TRANS-message is used to convey the initial NAS-message from the UE to the SRNC. Part of this message is the identification of the destination core network domain. If the initial NAS-message is destined for SAPI 3 (⇔ SS/SMS) and if the SRB 4 is established, then SRB 4 shall be used.

HO_UTRAN_FAIL
The UE shall send a HO_UTRAN_FAIL-message to the SRNC, if the hard handover to GSM failed or if the UE does not support handover to GSM. If the UE is unable to use the formerly established UTRA-physical channels, it shall perform a cell update procedure (cause: radio link failure) to obtain a channel allocation and then transfer the HO_UTRAN_FAIL-message.

MEAS_CTRL_FAIL
The MEAS_CTRL_FAIL-message is sent if the UE does not support the measurement configuration requested by the RNC.

Uplink DCCH-Messages

Message Type 5 bit / [bin]	RLC-Mode & Signal. Radio Bearer (SRB)	Messages on Uplink DCCH Name (ASN.1)	Abbreviation
01000	UM / AM (SRB 1 / 2)	measurementReport	MEAS_REP
01001	AM (SRB 2)	physicalChannelReconfigurationComplete	PHYS_CHAN_RECONF_COM
01010	AM (SRB 2)	physicalChannelReconfigurationFailure	PHYS_CHAN_RECONF_FAIL
01011	AM (SRB 2)	radioBearerReconfigurationComplete	RB_RECONF_COM
01100	AM (SRB 2)	radioBearerReconfigurationFailure	RB_RECONF_FAIL
01101	AM (SRB 2)	radioBearerReleaseComplete	RB_REL_COM
01110	AM (SRB 2)	radioBearerReleaseFailure	RB_REL_FAIL
01111	AM (SRB 2)	radioBearerSetupComplete	RB_SETUP_COM

Uplink DCCH-Messages

MEAS_REP
The MEAS_REP-message is used by the UE to report downlink measurement results and measurement results for measurements as ordered by the SRNC to the SRNC. These other measurements relate to traffic volume measurements etc.. The RLC-mode to be used (UM or AM) is signaled to the UE in the respective setup message (e.g. RRC_CONN_SETUP).

PHYS_CHAN_RECONF_COM
By sending the PHYS_CHAN_RECONF_COM-message the UE confirms the successful physical channel reconfiguration. This message shall also be used as response message for CELL_UPD_CNF in case of a cell update procedure when the SRNC reconfigures the physical channel configuration.

PHYS_CHAN_RECONF_FAIL
The PHYS_CHAN_RECONF_FAIL-message is sent by the UE on the old radio link to indicate that the reconfiguration of the physical channels did not succeed. In addition, the PHYS_CHAN_RECONF_FAIL-message is used by the UE to indicate to the SRNC that the allocated compressed mode radio configuration is inconsistent.

RB_RECONF_COM
By sending the RB_RECONF_COM-message the UE confirms the successful radio bearer reconfiguration. This message shall also be used as response message for CELL_UPD_CNF in case of a cell update procedure when the SRNC reconfigures the radio bearer configuration.

RB_RECONF_FAIL
The RB_RECONF_FAIL-message is sent by the UE on the old radio link to indicate that the reconfiguration of the radio bearers did not succeed.

RB_REL_COM
By sending the RB_REL_COM-message the UE confirms the successful radio bearer release. This message shall also be used as response message for CELL_UPD_CNF in case of a cell update procedure when the SRNC releases radio bearers.

RB_REL_FAIL
The RB_REL_FAIL-message is sent by the UE on the old radio link to indicate that the release of the radio bearer was not possible.

RB_SETUP_COM
By sending the RB_SETUP_COM-message the UE confirms the successful radio bearer setup.

Uplink DCCH-Messages

Message Type 5 bit / [bin]	RLC-Mode & Signal. Radio Bearer (SRB)	Messages on Uplink DCCH Name (ASN.1)	Abbreviation
10000	AM (SRB 2)	radioBearerSetupFailure	RB_SETUP_FAIL
10001	UM / AM (SRB 1 / 2)	rrcConnectionReleaseComplete	RRC_CONN_REL_COM
10010	AM (SRB 2)	rrcConnectionSetupComplete	RRC_CONN_SETUP_COM
10011	AM (SRB 2)	rrcStatus	RRC_STATUS
10100	AM (SRB 2)	securityModeComplete	SEC_MODE_COM
10101	AM (SRB 2)	securityModeFailure	SEC_MODE_FAIL
10110	AM (SRB 2)	signallingConnectionReleaseIndication	SIG_CONN_REL_IND
10111	AM (SRB 2)	transportChannelReconfigurationComplete	TrCH_RECONF_COM

Uplink DCCH-Messages

RB_SETUP_FAIL
The RB_SETUP_FAIL-message is sent by the UE on the old radio link to indicate that the setup of the radio bearer was not possible.

RRC_CONN_REL_COM
The RRC_CONN_REL_COM-message is sent by the UE to confirm the release of the RRC-connection. If sent in CELL_DCH-state, RLC-UM is used, in CELL_FACH, RLC-AM is used. If sent in CELL_DCH-state, the RRC_CONN_REL_COM-message is sent max. N308 times with a periodicity of T308 before physical channel release to increase the reception probability.

RRC_CONN_SETUP_COM
The RRC_CONN_SETUP_COM-message is sent by the UE to the SRNC to confirm the establishment of an RRC-connection.

RRC_STATUS
If the UE receives an RRC-message from the RNC with one or more protocol errors, it shall send an RRC-STATUS-message to the RNC indicating which received message caused the error and the error cause.

SEC_MODE_COM
By sending the SEC_MODE_COM-message the UE confirms the proper reception of the SEC_MODE_CMD-message and the start of the new or changed integrity check procedure.

SEC_MODE_FAIL
If the UE receives an invalid SEC_MODE_CMD-message or if the received security configuration is invalid, the UE shall send a SEC_MODE_FAIL-message to the SRNC, indicating the error reason.

SIG_CONN_REL_IND
The SIG_CONN_REL_IND-message is used by the UE to request for the release of one of its signaling connections to either core network domains. This procedure is similar to the network initiated signaling connection release procedure (see SIG_CONN_REL-message on downlink DCCH).

TrCH_RECONF_COM
By sending the TrCH_RECONF_COM-message the UE confirms the successful transport channel reconfiguration. This message shall also be used as response message for CELL_UPD_CNF in case of a cell update procedure when the SRNC reconfigures the transport channel configuration.

Uplink DCCH-Messages

Message Type 5 bit / [bin]	RLC-Mode & Signal. Radio Bearer (SRB)	Messages on Uplink DCCH Name (ASN.1)	Abbreviation
11000	AM (SRB 2)	transportChannelReconfigurationFailure	TrCH_RECONF_FAIL
11001	AM (SRB 2)	transportFormatCombinationControlFailure	TFC_CTRL_FAIL
11010	AM (SRB 2)	ueCapabilityInformation	UE_CAP_INFO
11011	AM (SRB 3 / 4)	uplinkDirectTransfer	UL_DIR_TRANS
11100	AM (SRB 2)	utranMobilityInformationConfirm	UTRAN_MOB_INFO_CNF
11101	AM (SRB 2)	utranMobilityInformationFailure	UTRAN_MOB_INFO_FAIL
11110	-/-	spare2	-/-
11111	-/-	spare1	-/-

Uplink DCCH-Messages

TrCH_RECONF_FAIL
The TrCH_RECONF_FAIL-message is sent by the UE on the old radio link to indicate that the reconfiguration of the transport channels did not succeed. For instance, the UE may not support the requested configuration.

TFC_CTRL_FAIL
The TFC_CTRL_FAIL-message is used by the UE to indicate to the SRNC, if a received TFC_CONTROL-message was invalid. Note: The UE shall only send the TFC_CTRL_FAIL-message, if the TFC_CTRL-message was sent by the network using SRB 2 (RLC-AM). If this message was sent on SRB 1 (RLC-UM), then the UE shall not send the TFC_CTRL_FAIL-message.
In either case, the UE shall continue using the former TFCS, if the received new configuration cannot be supported.

UE_CAP_INFO
The UE shall send a UE_CAP_INFO-message to the SRNC, whenever it receives a UE_CAP_ENQ-message from the SRNC and when the UE's radio access or inter-RAT-capabilities (GSM, cdma2000) change.

UL_DIR_TRANS
The UL_DIR_TRANS-message is used by the UE for the embedded transfer of NAS-messages towards the SRNC. Note that the first NAS-message to be sent on a new signaling connection will be rather embedded into an INIT_DIR_TRANS-message.

UTRAN_MOB_INFO_CNF
The UE shall send a UTRAN_MOB_INFO_CNF-message in case of a cell update or URA update procedure e.g. when it receives new U-RNTI- and/or C-RNTI-values from the SRNC. Note that other messages like TrCH_RECONF_COM shall be used as response message to UTRAN_MOB_INFO when the SRNC reconfigures the radio bearer, TrCH or physical channel configuration.

UTRAN_MOB_INFO_FAIL
The UE shall send a UTRAN_MOB_INFO_FAIL-message in case of a cell update or URA update procedure after reception of an invalid security configuration in the UTRAN_MOB_INFO-message.

Note: Spare-values are reserved for future use.

Downlink CCCH-Messages

Message Type 3 bit / [bin]	RLC-Mode & Signal. Radio Bearer (SRB)	Messages on Downlink CCCH Name (ASN.1)	Abbreviation
000	UM (SRB 0) *)	cellUpdateConfirm	CELL_UPD_CNF *)
001	UM (SRB 0)	rrcConnectionReject	RRC_CONN_REJ
010	UM (SRB 0) *)	rrcConnectionRelease	RRC_CONN_REL *)
011	UM (SRB 0)	rrcConnectionSetup	RRC_CONN_SETUP
100	UM (SRB 0) *)	uraUpdateConfirm	URA_UPD_CNF *)
101	-/-	spare3	-/-
110	-/-	spare2	-/-
111	-/-	spare1	-/-

*) Message is also defined for downlink DCCH

Downlink CCCH-Messages

CELL_UPD_CNF
The CELL_UPD_CNF-message is sent by the SRNC on CCCH to confirm a cell update of a UE. This message can only be sent on CCCH, if the cell update was not accompanied by an SRNS-relocation and if ciphering shall not be performed.

RRC_CONN_REJ
The RRC_CONN_REJ-message is sent by the RNC to reject the UE's attempt to establish an RRC-connection. Possible reason is overload in the cell. Accordingly, the RNC will direct the UE to another cell as part of the RRC_CONN_REJ-message.

RRC_CONN_REL
The RNC will send an RRC_CONN_REL-message to release an RRC-connection of a UE. This message may only be sent on CCCH, if the UE is in CELL_FACH-state and the DCCH is currently not available in UTRAN.

RRC_CONN_SETUP
The RRC_CONN_SETUP-message is sent by the RNC to confirm the RRC-connection establishment and to configure the respective signaling radio bearers.

URA_UPD_CNF
The URA_UPD_CNF-message is sent by the SRNC on CCCH to confirm an URA update of a UE. This message can only be sent on CCCH, if the URA update was not accompanied by an SRNS-relocation. Even if no SRNS-relocation was performed, the SRNC may still send this message on DCCH.

Note: Spare-values are reserved for future use.

Uplink CCCH-Messages

Message Type 2 bit / [bin]	RLC-Mode & Signal. Radio Bearer (SRB)	Messages on Uplink CCCH Name (ASN.1)	Abbreviation
00	TM (SRB 0)	cellUpdate	CELL_UPD
01	TM (SRB 0)	rrcConnectionRequest	RRC_CONN_REQ
10	TM (SRB 0)	uraUpdate	URA_UPD
11	-/-	spare	-/-

Uplink CCCH-Messages

CELL_UPD
The CELL_UPD-message is sent by the UE to initiate the cell update procedure. Note that the cell update procedure is used not only in case of periodical updating or cell reselection but also as a means to indicate an RLC-unrecoverable error or for paging response purposes.

RRC_CONN_REQ
The UE will send an RRC_CONN_REQ-message to initiate the RRC-connection establishment procedure.

URA_UPD
The URA_UPD-message is used by the UE in case of cell reselection to a cell which belongs to a different URA than the previous serving cell and if the UE is in URA_PCH-state.

Note: Spare-values are reserved for future use.

PCCH- (through PCH) and BCCH (through FACH)-Messages

Message Type 1 bit / [bin]	RLC-Mode & Signal. Radio Bearer (SRB)	Messages on PCCH Name (ASN.1)	Abbreviation
0	TM (no SRB / PCH)	pagingType1	PAG_TYPE1
1	-/-	spare	-/-
Message Type 2 bit / [bin]	**RLC-Mode & Signal. Radio Bearer (SRB)**	**Messages on BCCH Name (ASN.1)**	**Abbreviation**
00	TM (no SRB / FACH)	systemInformation	SYS_INFO
01	TM (no SRB / FACH)	systemInformationChangeIndication	SYS_INFO_CHANGE_IND
10	-/-	spare2	-/-
11	-/-	spare1	-/-

PCCH- (through PCH) and BCCH (through FACH)-Messages

PAG_TYPE1
The PAG_TYPE1-message is used in RRC-idle mode, CELL_PCH- and URA_PCH-mode to convey either core network's paging request to the UE. If the UE is in RRC-idle mode, an RRC-connection needs to be established to send a suitable paging response to the paging core network domain. If the UE is in CELL_PCH- or URA_PCH-state, the UE shall perform a cell update, thereby move to CELL_FACH-state and then transmit a suitable paging response message to the paging core network domain.

SYS_INFO
The UTRAN shall broadcast SIB10 embedded in SYS_INFO-messages over FACH for UE's in CELL_DCH-state, if the UE's support the simultaneous reception of one DPCH-physical channel and the S-CCPCH which carries the FACH.

SYS_INFO_CHANGE_IND
The CRNC uses the SYS_INFO_CHANGE_IND-message to inform UE's in CELL_FACH-state about an upcoming adjustment of the system information in that cell.

Note: Spare-values are reserved for future use.

UMTS – Signaling & Protocol Analysis

Measurements on the UE-Side

Measurement Type	Reporting Event ID's
Intra-Frequency	FDD: 1A – 1F / TDD: 1G – 1I
Inter-Frequency	2A – 2F
Inter-RAT	3A – 3D
Traffic Volume	4A – 4B
Quality	5A
UE Internal	6A – 6G
UE Positioning	7A – 7C

UMTS – Signaling & Protocol Analysis

Measurements on the UE-Side

In UMTS, various different measurement types can be defined by the SRNC (⇔ for the UE) and the CRNC (⇔ for the NodeB).

- **Measurements on the UE-Side**
  While in RRC-idle mode, the RRC: SIB 11 determines which measurements shall be conducted by the UE. In RRC-connected mode, the measurement procedures of the UE are additionally controlled by the RRC: MEAS_CTRL-message and possibly by the RRC: SIB 12. As the table on the graphics page indicates, there are various measurement possibilities defined for the UE. The UE shall send MEAS_REP-messages either periodically or event triggered (⇔ default). Each possible event is numbered (1A, …, 7C) and can be selected by the RNC for setting up the UE.

  ⇒ **Intra-Frequency Measurements**
  These measurements relate to code signal measurements or (code signal vs. overall frequency signal)-measurements on the same UTRA-frequency by the UE

  ⇒ **Inter-Frequency Measurements**
  Same as for Intra-frequency measurements but on different UTRA-carriers (FDD and/or TDD).

  ⇒ **Inter-RAT-Measurements**
  These measurements relate most likely to received signal and BSIC-measurements on GSM-frequency bands.

  ⇒ **Traffic Volume-Measurements**
  Traffic volume measurements provide means to measure the load on dedicated and common TrCH's to provide decision options for TrCH-reconfiguration. Traffic Volume Measurements have been dealt with during the presentation of the MAC-layer.

  ⇒ **Quality-Measurements**
  Quality measurements relate to the measurement of the bit error rate on TrCH's (based on pilot bits in DPDCH/D) and/or physical channels (based on pilot bits on DPCCH/D).

  ⇒ **UE Internal-Measurements**
  UE internal measurements relate to the supervision of the minimum and maximum UE output power and similar UE RF-parameters.

  ⇒ **UE-Positioning-Measurements**
  These measurements are related to location based services.

- **Measurements on the NodeB-Side**
  The measurements to be conducted by the NodeB are controlled through NBAP: COMM_MEAS_INIT-messages for common TrCH's and NBAP: DEDIC_MEAS_INIT-messages for dedicated TrCH's.

[3GTS 25.331 (8.4), (14.1 – 14.7)]

* Larger Transport format mean lower spreading code meaning high Tx power

Example of a MEAS_CTRL-Message (Part 1)

```
+---------+-----------------------------------------+---
|1.1 measurementControl
|1.1.1 r3
|1.1.1.1 measurementControl-r3
|---00---  |1.1.1.1.1 rrc-TransactionIdentifier       |0
|***b4***  |1.1.1.1.2 measurementIdentity             |9
|1.1.1.1.3 measurementCommand
|1.1.1.1.3.1 setup
|1.1.1.1.3.1.1 intraFrequencyMeasurement
|1.1.1.1.3.1.1.1 intraFreqCellInfoList
|1.1.1.1.3.1.1.1.1 newIntraFreqCellList
 1.1.1.1.3.1.1.1.1.1 newIntraFreqCell
 ***b5***  |1.1.1.1.3.1.1.1.1.1.1 intraFreqCellID        |0
 1.1.1.1.3.1.1.1.1.1.2 cellInfo
 ***b6***  |1.1.1.1.3.1.1.1.1.1.2.1 cellIndividualOffset |-20
 1.1.1.1.3.1.1.1.1.1.2.2 modeSpecificInfo
 1.1.1.1.3.1.1.1.1.1.2.2.1 fdd
 1.1.1.1.3.1.1.1.1.1.2.2.1.1 primaryCPICH-Info
 ***b9***  |1.1.1.1.3.1.1.1.1.1.2.2.1.1.1 primaryScramb..|451
 ***b6***  |1.1.1.1.3.1.1.1.1.1.2.2.1.2 primaryCPICH-TX..|30
 ---1----  |1.1.1.1.3.1.1.1.1.1.2.2.1.3 readSFN-Indicator|1
 ----0---  |1.1.1.1.3.1.1.1.1.1.2.2.1.4 tx-DiversityInd..|0
```

Setup, Modify and Release are possible.

Dynamic reference number of this measurement type (intra-frequency) to be used by the UE in the respective MEAS_REP's and the RNC for upcoming MEAS_CTRL-messages.

This MEAS_CTRL-message sets up intra-frequency measurements

Specification of one or more intra-frequency NC's (the presented cell only illustrates one out of 6 cells that have been configured in this message.

MS should be able to handle up to 32 neighbours cells on the same freq. channel

Example of a MEAS_CTRL-Message (Part 1)

Intentionally left blank

© INACON GmbH 2003 - 2005. All rights reserved. Reproduction and/or unauthorized use of this material is prohibited and will be prosecuted to the full extent of German and international laws. Version Number: 1.6

Example of a MEAS_CTRL-Message (Part 2)

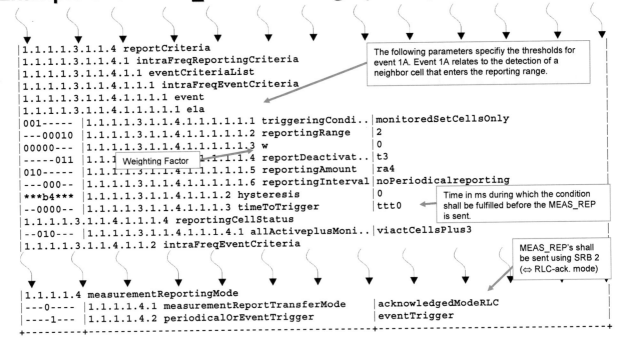

Example of a MEAS_CTRL-Message (Part 2)

Note: The measurement quantity in the presented example was CPICH Ec/No.

Important Measurement Parameters

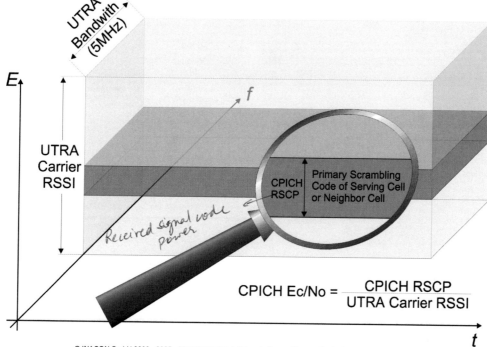

$$\text{CPICH Ec/No} = \frac{\text{CPICH RSCP}}{\text{UTRA Carrier RSSI}}$$

Important Measurement Parameters

UTRA Carrier RSSI
The UTRA-Carrier RSSI (Received Signal Strength Indicator) is a parameter in dB_m that describes the total signal strength of a UTRA-carrier frequency. Accordingly, the UTRA-Carrier RSSI represents the receivable energy of all cells at a certain location. [3GTS 25.215 (5.1.3)]

UTRA-Carrier RSSI / Range: 0 – 76 (7 bit)	Value [dB_m] / Step-Size: 1 dB_m
0	Value < -100
1 – 75	-(101 – UTRA-Carrier-RSSI) ≤ Value < -(100 – UTRA-Carrier-RSSI)
76	Value ≥ -25

CPICH RSCP
The CPICH RSCP (Received Signal Code Power) is a parameter in dB_m that describes the signal strength of the P-CPICH of any given cell. Note that the spreading code of the P-CPICH is identical for all cells while the primary scrambling code is different. [3GTS 25.215 (5.1.1)]

CPICH-RSCP / Range: 0 – 91 (7 bit)	Value [dB_m] / Step-Size: 1 dB_m
0	Value < -115
1 – 90	-(116 – CPICH-RSCP) ≤ Value < -(115 – CPICH-RSCP)
91	Value ≥ -25

CPICH Ec/No
The CPICH Ec/No (Energy per chip vs. total power) is a relative parameter in dB that sets the CPICH RSCP in relation to UTRA-carrier RSSI (⇒ CPICH Ec/No = (CPICH RSCP / UTRA-Carrier RSSI). [3GTS 25.215 (5.1.5)]

CPICH-Ec-No / Range: 0 – 49 (6 bit)	Value [dB] / Step-Size: 0.5 db
0	Value < -24
1 – 48	-(24.5 – (CPICH-Ec-No / 2)) ≤ Value < -(24 – (CPICH-Ec-No / 2))
49	Value ≥ 0

Bit Error Rate on TrCH and Physical Channel

Phy. channel → - pilot bits (predetermined bit seq.)
- power info.

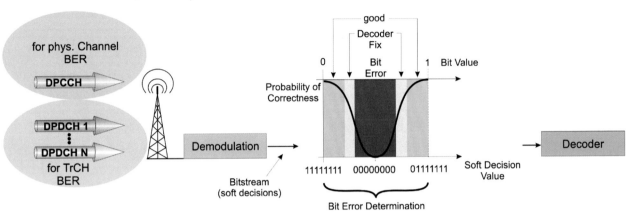

Bit Error Rate on TrCH and Physical Channel

⇒ BER-measurements in UMTS are performed by the receiver immediately before the channel decoding process. The BER-measurements shall be performed on all non-punctured bits.
⇒ The decision to declare a bit error is based on the "soft decision" bits which are evaluated during or after the demodulation process.
⇒ The accuracy of the BER-measurement shall be ≤ 10% if the BER ≤ 30%. For higher BER, the accuracy is undefined.
⇒ The NodeB shall measure the BER of all information bits on DPCCH *or* DPDCH within one TTI and shall provide the result to the SRNC as QE-value.
⇒ Whether the BER on physical channels or on transport channels is measured and reported is configured by the SRNC for uplink measurements by the NodeB during the radio link setup procedure (NBAP: INIT_MESS / RL_SETUP-message / IE: QE-selector).

- **BER on Physical Channels**
  For physical channel BER-measurement, all information bits on the DPCCH of an active link set are used.

- **BER on Transport Channels**
  For TrCH BER-measurement, all non-punctured information bits on all DPDCH's of the selected TrCH are used.

The following table provides some key values for the interpretation of QE-values in recording files:

Result of the Measurement →

QE-Value	app. Bit Error Rate (BER)
0 … 8	< 1%
58	≈ 2.5%
100	≈ 5.5%
132	≈ 10%
169	≈ 20%
191	≈ 30%
206	≈ 40%

QE-Value	app. Bit Error Rate (BER)
218	≈ 50%
228	≈ 60%
236	≈ 70%
243	≈ 80%
249	≈ 90%
255	100%

[3GTS 25.215 (5.2.6, 5.2.7), 3GTS 25.133 (9.2.7.2, 9.2.9.2)]

Example: QE in Iub-FP: UL-DATA represents the BER

```
+---------+-----------------------------------------------+-------------------------------------+
|FP Data Frame DCH                                                                              |
|                                                                                               |
|         |2.1.7 RLC: Acknowledgement Super Field                                               |
|0010---- |2.1.7.1 RLC: SUFI Type                         |Acknowledgement                      |
|**b12*** |2.1.7.2 RLC: Last Sequence Number              |2                                    |
|2.1.8 RLC: Padding                                        QE = 100                             |
|**b100** |2.1.8.1 RLC: Padding                            ⇒                                    |
|                                                          BER phys. Channel or BER TrCH ≈ 5.5 % LM(GM)
|                                                         1010101010101010101010101010101010 B  |
|2.2 FP: Padding                                                                                |
|----0000 |2.2.1 FP: Padding                              |0                                    |
|3 FP: Trailer                                                                                  |
|11001000 |3.1 FP:  Quality Estimate                      |100                                  |
|0------- |3.2 FP:  CRC Indicator (Transport Block)       |Correct                              |
|-0000000 |3.3 FP:  Padding                               |0                                    |
|***B2*** |3.4 FP:  Payload CRC                           |'4144'H                              |
+---------+-----------------------------------------------+-------------------------------------+
```

Example: QE in Iub-FP: UL-DATA represents the BER

The illustrated Iub-FP: UL-DATA-PDU contains one or more transport blocks plus a piggybacked RLC-STATUS-PDU (⇔ ACK-SUFI). In its trailer, each Iub-FP: UL-DATA-PDU also provides the Quality Estimate (QE) which stems from the physical layer in the NodeB.
In the example, the QE-value = 100_{dec} which represents a fairly poor quality and a bit error rate of app. 5.5%.

UMTS – Signaling & Protocol Analysis

- *A Comprehensive Inside View on UMTS*
- *Signaling & Protocol Analysis in UTRAN*
- **Important UMTS Scenarios & Call Tracing**

UMTS – Signaling & Protocol Analysis

Intentionally left blank

Explanations of the Used Message Descriptors

- ### Message Descriptors on Uu-Interface

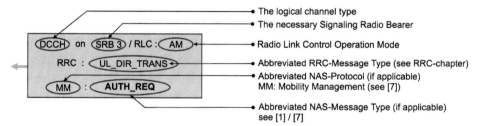

- The logical channel type
- The necessary Signaling Radio Bearer
- Radio Link Control Operation Mode
- Abbreviated RRC-Message Type (see RRC-chapter)
- Abbreviated NAS-Protocol (if applicable) MM: Mobility Management (see [7])
- Abbreviated NAS-Message Type (if applicable) see [1] / [7]

- ### Message Descriptors on Iu-cs-, Iu-ps- and Iur-Interface

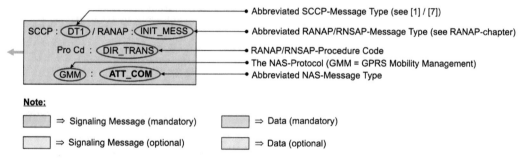

- Abbreviated SCCP-Message Type (see [1] / [7])
- Abbreviated RANAP/RNSAP-Message Type (see RANAP-chapter)
- RANAP/RNSAP-Procedure Code
- The NAS-Protocol (GMM = GPRS Mobility Management)
- Abbreviated NAS-Message Type

Note:

- ⇒ Signaling Message (mandatory)
- ⇒ Data (mandatory)
- ⇒ Signaling Message (optional)
- ⇒ Data (optional)

Explanations of the Used Message Descriptors

Note:
- For layout reasons, we use abbreviations for all presented messages and data PDU's. These abbreviations are consistent within the INACON documentation set.
- All UMTS specific abbreviations (e.g. RRC-protocol, RLC-protocol, RANAP-protocol, ..) are explained in this document.
- For other abbreviations, please refer to [1], [2], [5] and [7].
- Different colors are used to distinguish data PDU's from control PDU's and mandatory from optional PDU's.
- The most important information in a PDU is presented in broad letters.

Message Descriptors on Uu-Interface
On Uu-interface, all information is transferred in logical channels (DCCH, DTCH, ...). Mostly, signaling radio bearers (SRB) are used for the transfer of control plane information. In any case, all information is transmitted in RLC-PDU's. Whether or not higher layer PDU's are included, depends on the very scenario.

Message Descriptors on Iu-cs-, Iu-ps- and Iur-Interface
The illustrated message descriptors only relate to messages within the control plane. Whether or not NAS-content is included in a message, depends on the message itself. Sometimes, the RNSAP/RANAP-procedure code is the most important content.

Note: ALCAP-messages on Iu-cs- and Iur-interface are formatted as on the Iub-interface and are explained on the following page. On Iur-interface (and with Rel. 4 also on Iu-cs-interface) ALCAP-messages can be sent over an IP-based protocol stack.

Message Descriptors on Iub-Interface

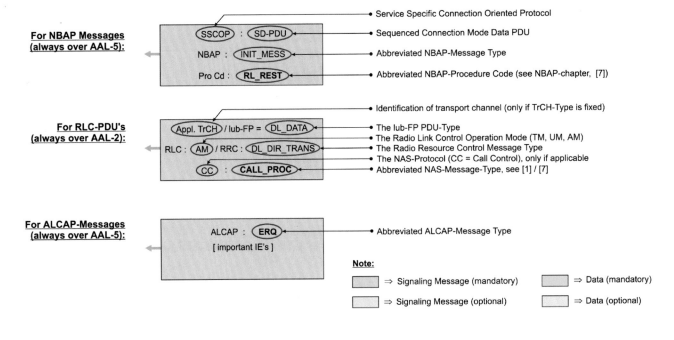

Message Descriptors on Iub-Interface

On Iub-interface, three different protocol groups have to be distinguished:

NBAP-Messages (always over AAL-5)
NBAP-messages are always transmitted over the SSCOP-protocol and use SD-PDU's as bearers. The most important content is the NBAP-procedure code.

RLC-PDU's (always over AAL-2)
RLC-PDU's are always transmitted over AAL-2-links. Important to know are the used transport channel, the Iub-FP-PDU-type, the RLC-mode and, if applicable, the embedded RRC- and NAS-message types.
Note: We use the same message format for the presentation of stand-alone Iub-FP PDU's (e.g. DL-SYNC).

ALCAP-Messages (over AAL-5)
ALCAP-message are part of the transport network control plane. Different options exist on many interfaces for the underlying transport network. We present the ALCAP-message type plus the most important message content.

Common Scenarios
- **NodeB Setup**

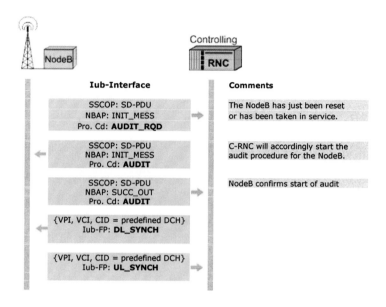

Common Scenarios

NodeB Setup

NodeB Setup

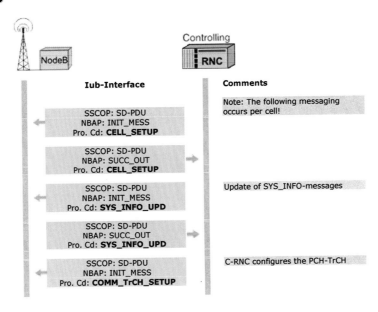

NodeB Setup

NodeB Setup

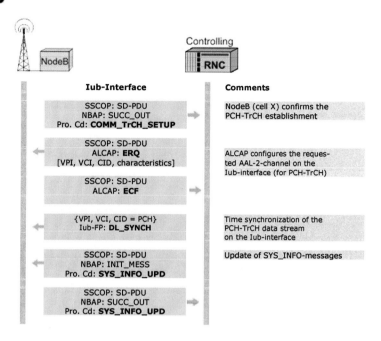

NodeB Setup

NodeB Setup

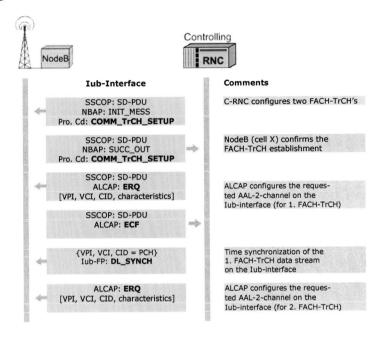

NodeB Setup

NodeB Setup

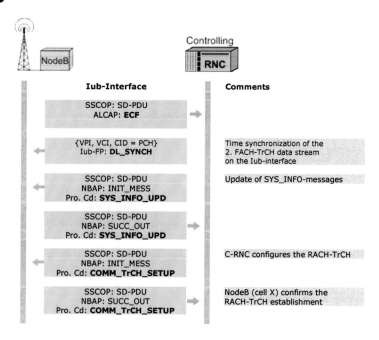

NodeB Setup

NodeB Setup

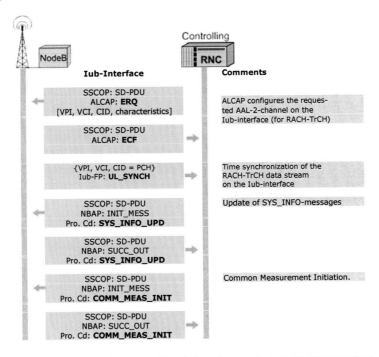

NodeB Setup

Circuit-Switched Scenarios

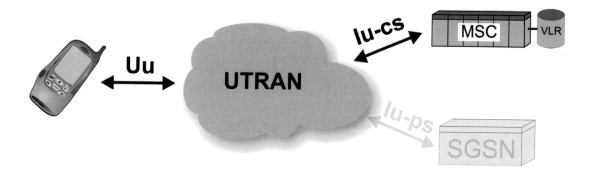

Circuit-Switched Scenarios

Circuit-switched procedures include all procedures that relate to the UE communicating with the circuit-switched core network domain through the Iu-cs-interface.

(1) UE-Registration

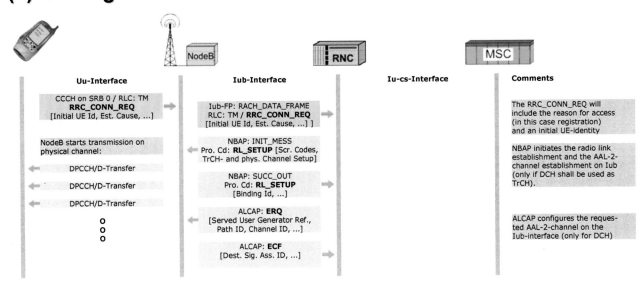

(1) UE-Registration

Initial Conditions
The UE is in RRC-idle mode and shall return to RRC-idle mode after successful completion of the procedure. The network operates in NOM II (⇔ no Gs-interface). Therefore the registration procedures are executed in non-combined fashion.

Note: If the Gs-interface would be present and if the network would operate in NOM I, a mobile station which can operate both circuit-switched and packet-switched applications, can choose to perform circuit-switched and packet-switched registration procedures in combined fashion [3GTS 23.060 (6.3.4.1)].

Applicability of this Procedure
This procedure is applicable in case of IMSI attachment, periodic location updating and normal location updating. The very reason for the registration procedure is indicated within the LOC_UPD_REQ-message.

Description
⇒ The UE will start the procedure by sending an RRC_CONN_REQ-message to the RNC.

Note: The following procedure is optional. It will be conducted,
- If the RNC wants to perform the procedure using a DCH-TrCH in uplink and downlink direction.

⇒ The RNC will send an NBAP: RL_SETUP-message (initiating message) to the NodeB to request for the establishment of an AAL-2 channel for the DCH-TrCH on the Iub-interface. The NBAP: RL_SETUP-message also fully specifies the new DCCH's, TrCH's and physical channels in uplink and downlink direction. The NodeB will configure the requested physical resources in uplink and downlink direction and will already start to transmit in downlink direction on the new DPCCH/D. Then the NodeB will send an NBAP: RL_SETUP-message (successful outcome) back to the CRNC. This message contains the Binding Id., to be used by the CRNC as "Served User Generator Reference" to link the ALCAP: ERQ-message to the respective radio link setup procedure.
⇒ After positive confirmation of the NodeB, the RNC will send an ALCAP: ERQ-message to the NodeB to physically allocate the AAL-2 channel. This message contains an identification of the AAL-2 Path ID and the AAL-2 Channel ID that shall be used for the new DCH-TrCH. The NodeB confirms the AAL-2 channel establishment by replying an ALCAP: ECF-message. The ALCAP: ERQ- and ECF-messages are related to each other through the Originating and Destination Signaling Association ID's.

(2) UE-Registration

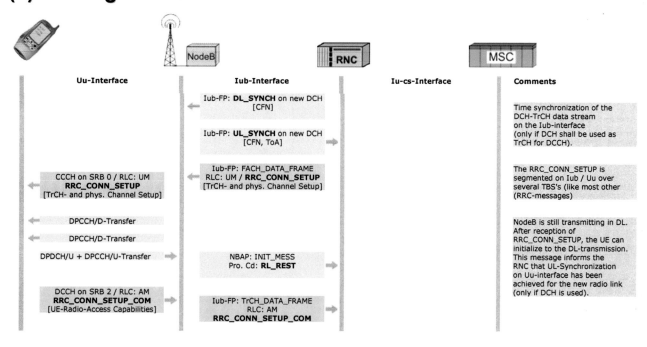

(2) UE-Registration

Description

⇒ The RNC-will start the DCH-TrCH-synchronization by sending an Iub-FP: DL_SYNCH-message to the NodeB. Synchronization is achieved as soon as the NodeB responds with an Iub-FP: UL_SYNCH-message. Finally, the DCH-TrCH is configured. Note that the presented messages are only there, if a DCH-TrCH shall be used for the registration process.

⇒ On the applicable TrCH (FACH), the RNC will convey an RRC_CONN_SETUP-message to the UE. This message will, among other things, configure the signaling radio bearers on DCCH 1 – N plus it will tell the UE whether to move into RRC / CELL_FACH-state (⇔ if common TrCH's shall be used) or into RRC / CELL_DCH-state (⇔ if dedicated TrCH's shall be used).

⇒ If DCH's are used, the NodeB will achieve physical channel synchronization on the new DPCH with the UE. This is indicated towards the RNC by transmitting an RL_REST-message.

⇒ If common TrCH's are used, the UE will confirm the RRC-connection establishment by transmitting an RRC_CONN_SETUP_COM-message on CPCH- or RACH-TrCH towards the RNC. Otherwise the uplink DCH-TrCH is used.

(3) UE-Registration

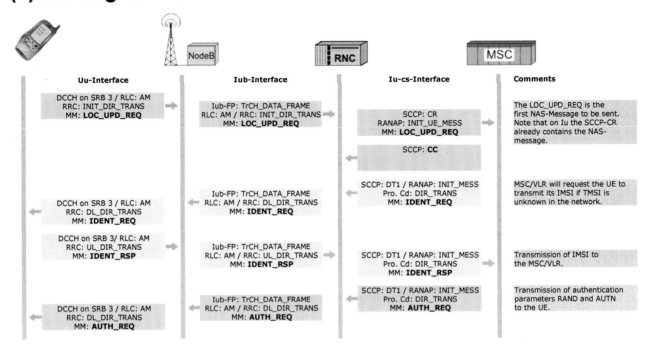

(3) UE-Registration

Description

⇒ The UE will use the DCCH on SRB 3 to transmit the LOC_UPD_REQ-message towards the RNC, embedded in an RRC: INIT_DIR_TRANS-message. The RNC will forward the LOC_UPD_REQ-message towards the MSC/VLR.

⇒ Note that an SCCP-connection is required to perform the registration procedure. Besides, RANAP requires an INIT_UE_MESS, to establish an Iu signaling connection between the MSC/VLR and the RNC. Accordingly, the LOC_UPD_REQ-message will be embedded into a RANAP: INIT_UE_MESS which in turn will be embedded into an SCCP: CR-message. The MSC/VLR will confirm SCCP-connection establishment by sending an SCCP: CC-message to the RNC. Optionally, the SCCP: CR- and CC-messages for connection establishment may be exchanged prior to transmitting the INIT_UE_MESS [LOC_UPD_REQ]-message in an SCCP: DT1-message.

Note: The following section is optional. It will be conducted:
- if the UE has *not* identified itself in LOC_UPD_REQ through its IMSI.

and
- if the MSC/VLR *cannot* identify the UE based on its TMSI (neither in the VLR-database nor by possibly invoking the UE's authentication information from the previous VLR.

⇒ The MSC/VLR will request the UE to transmit its IMSI by sending an IDENT_REQ-message to the UE.
⇒ The UE will reply by sending its IMSI to the MSC/VLR in an IDENT_RSP-message.

⇒ If necessary, the MSC/VLR will invoke the UE's authentication quintet from the HLR or from the previous VLR, before the AUTH_REQ-message is sent to the UE.

(4) UE-Registration

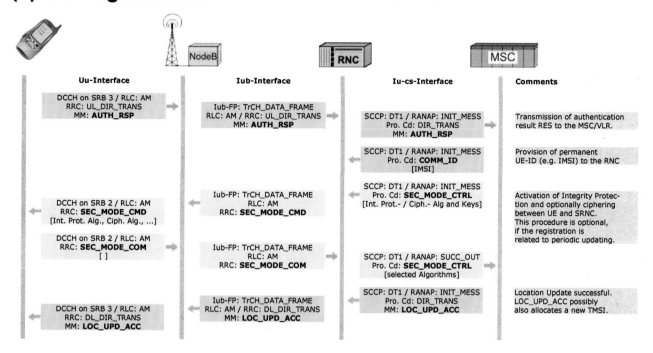

(4) UE-Registration

Description
⇒ The UE will calculate the authentication result RES and transmit it back the MSC/VLR.
⇒ At this time, the MSC will provide the permanent NAS (Non-Access-Stratum) UE-identity to the RNC (e.g. IMSI).

> **Note:**
> - The following part is optional, if the registration procedure is related to periodic location updating.
> - In this case, security related parameters are already known in the UE, in UTRAN and in the MSC/VLR.

⇒ The MSC/VLR will use the RANAP: SEC_MODE_CTRL-message to convey setup information related to ciphering and integrity protection towards the RNC. In turn, the RNC will send an RRC: SEC_MODE_CMD-message towards the UE to start the integrity protection procedure and optionally encryption.
⇒ The UE shall confirm start of the integrity protection procedure and of ciphering by sending an RRC: SEC_MODE_COM-message to the RNC.
⇒ The RNC confirms the successful outcome of the RANAP-security mode control procedure by sending the RANAP: SEC_MODE_CTRL-message to the MSC/VLR.

Finally, the MSC/VLR will send the LOC_UPD_ACC-message to the UE. This message confirms the registration of the UE towards the circuit-switched core network and may also contain a new TMSI.

(5) UE-Registration

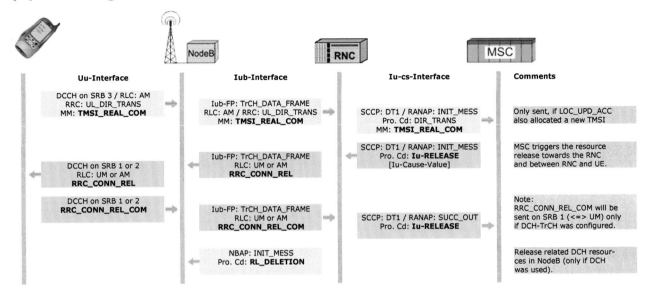

(5) UE-Registration

Description
⇒ Only if the LOC_UPD_ACC-message allocated a new TMSI to the UE, the UE shall confirm the new TMSI by sending a TMDI_REAL_COM-message to the MSC/VLR.
⇒ The MSC will trigger the resource release on all interfaces by sending an Iu-RELEASE-message to the RNC.
⇒ Consequently, the RNC will send one or more RRC_CONN_REL-message to the UE. If the UE is in CELL_FACH-state, then this message will be sent using RLC-AM-operation mode (⇔ SRB 2). In CELL_DCH-state, RLC-UM-operation mode shall be used (⇔ SRB 1).
⇒ The UE confirms release of he RRC-connection and state change back to RRC-idle by sending an RRC_CONN_REL_COM-message to the RNC. As the comment on the graphics page indicates, the RRC_CONN_REL_COM-message will be sent using SRB 1 (⇔ RLC-UMD-PDU's), if the UE is in CELL_DCH state, that is, when there are DCH's configured. Alternatively, if the UE is in CELL_FACH state (⇔ no DCH's are configured), the UE will use SRB 2 and RLC-AMD-PDU's. If the UE is in CELL_DCH-state, the UE shall retransmit the RRC_CONN_REL_COM-message up to N308 times until the physical and transport channel resources are released.
⇒ Finally, the RNC will confirm the resource release by sending Iu-RELEASE (successful outcome) back the MSC/VLR.

Note: The following procedure is optional. It will be conducted,
- If the RNC had allocated DCH-TrCH's in uplink and downlink direction.

⇒ The RNC will send an RL_DELETION-message to the NodeB to initiate the release of the dedicated TrCH and of the physical resources between NodeB and UE.

(6) UE-Registration

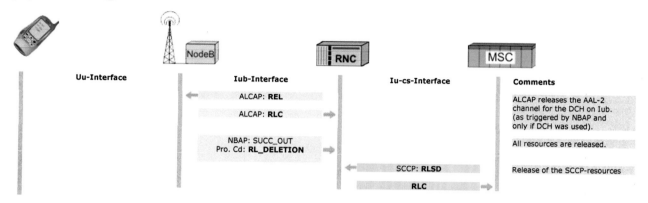

(6) UE-Registration

Description

⇒ Through the ALCAP-protocol (REL-message), the RNC will initiate the release of the related AAL-2 channel on the Iub-interface.
⇒ The ALCAP-protocol within the NodeB will confirm the release of the AAL-2 channel by sending a related RLC-message back to the RNC.
⇒ The NodeB will confirm the release of the dedicated TrCH and of the physical resources by sending an RL_DELETION-message (successful outcome) back the RNC.

⇒ The SCCP-connection is released by the MSC by sending an RLSD-message to the RNC. The RNC-confirms by sending an RLC-message back to the MSC. The SCCP-connection release will most likely occur already after the reception of the Iu-RELEASE-message (successful outcome) by the MSC.

(1) Mobile Originating Conversational Call

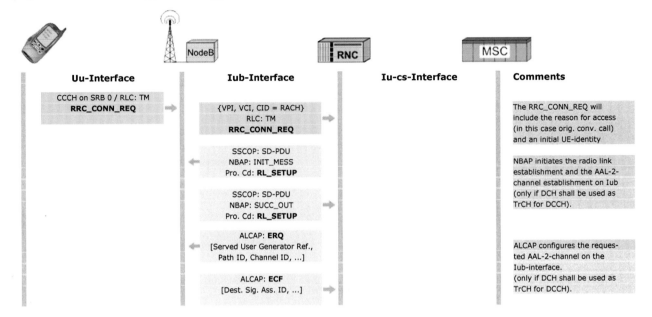

(1) Mobile Originating Conversational Call

Initial Conditions
The UE is in RRC-idle mode and shall return to RRC-idle mode after successful completion of the procedure. The UE receives the request from the user to establish a call.

Applicability of this Procedure
This procedure is applicable in case mobile originating call establishment towards the circuit-switched core network domain.

Description
⇒ The UE will start the procedure by sending RRC_CONN_REQ to the RNC.

> **Note:** The following procedure is optional. It will be conducted,
> • If the RNC wants to perform the procedure using a DCH-TrCH in uplink and downlink direction.

⇒ The RNC will send an RL_SETUP-message to the NodeB to request for the establishment of an AAL-2 channel for the DCH-TrCH on the Iub-interface. The NodeB will configure the requested physical resources in uplink and downlink direction and send a NBAP: RL_SETUP-message (successful outcome) back to the CRNC. This message also contains the Binding Id., to be used by the CRNC as "Served User Generator Reference" to link the ALCAP: ERQ-message to the respective RL_SETUP-message.
⇒ After positive confirmation of the NodeB, the RNC will send an ALCAP: ERQ-message to the NodeB to physically allocate the AAL-2 channel.
⇒ The NodeB confirms the AAL-2 channel establishment by replying with an ALCAP: ECF-message.

(2) Mobile Originating Conversational Call

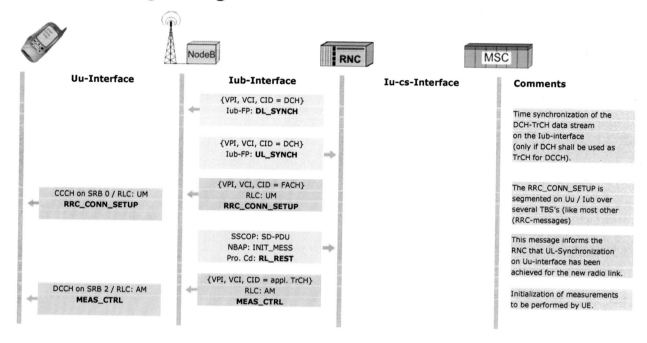

(2) Mobile Originating Conversational Call

Description

⇒ The RNC will start the TrCH-synchronization by sending an Iub-FP: DL_SYNCH-message to the NodeB. Synchronization is achieved as soon as the NodeB responds with an Iub-FP: UL_SYNCH-message. Finally, the DCH-TrCH is configured. Note that the presented messages are only there, if a DCH-TrCH shall be used for the registration process.

⇒ On the applicable TrCH (FACH), the RNC will convey an RRC_CONN_SETUP-message to the UE. This message will, among other things, configure the signaling radio bearers on DCCH 1 – N plus it will tell the UE whether to move into RRC / CELL_FACH-state (⇔ if common TrCH's shall be used) or into RRC / CELL_DCH-state (⇔ if dedicated TrCH's shall be used).

⇒ If DCH's are used, the NodeB will now achieve physical channel synchronization on the new DPCH with the UE. This is indicated towards the RNC by transmitting an RL_REST-message.

⇒ The SRNC will transmit a MEAS_CTRL-message to the UE indicating to the UE which measurements shall be performed.

⇒ If common TrCH's are used, the UE will confirm the RRC-connection establishment by transmitting an RRC_CONN_SETUP_COM-message on CPCH- or RACH-TrCH towards the RNC. Otherwise the uplink DCH-TrCH is used.

(3) Mobile Originating Conversational Call

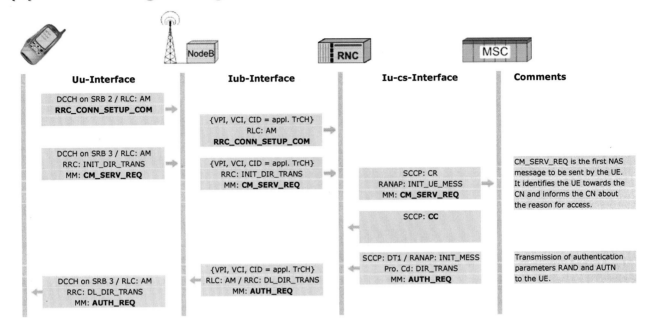

(3) Mobile Originating Conversational Call

Description
⇒ If common TrCH's are used, the UE will confirm the RRC-connection establishment by transmitting an RRC_CONN_SETUP_COM-message on CPCH- or RACH-TrCH towards the RNC. Otherwise the uplink DCH-TrCH is used.
⇒ The UE will use the DCCH on SRB 3 to transmit the CM_SERV_REQ-message towards the RNC, embedded in an RRC: INIT_DIR_TRANS-message. The RNC will forward the CM_SERV_REQ-message towards the MSC/VLR.
⇒ Note that an SCCP-connection is required to perform the registration procedure. Besides, RANAP requires an INIT_UE_MESS, to establish an Iu signaling connection between the MSC/VLR and the RNC. Accordingly, the CM_SERV_REQ -message will be embedded into a RANAP: INIT_UE_MESS which in turn will be embedded into an SCCP: CR-message. The MSC/VLR will confirm SCCP-connection establishment by sending an SCCP: CC-message to the RNC. Optionally, the SCCP: CR- and CC-messages for connection establishment may be exchanged prior to transmitting the INIT_UE_MESS [CM_SERV_REQ]-message in an SCCP: DT1-message.
⇒ The MSC/VLR will initiate the user authentication by sending an AUTH_REQ-message to the UE.

(4) Mobile Originating Conversational Call

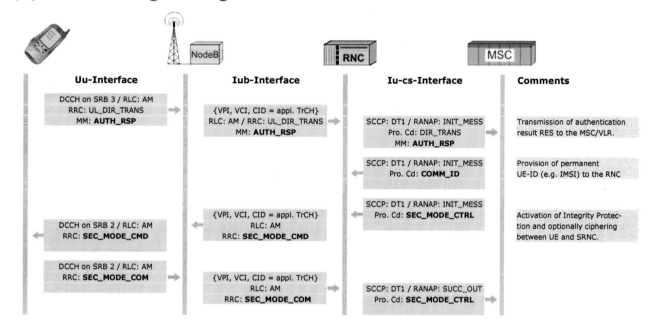

(4) Mobile Originating Conversational Call

Description

⇒ The UE will calculate the authentication result RES and transmit it back the MSC/VLR.
⇒ At this time, the MSC will provide the permanent NAS (Non-Access-Stratum) UE-identity to the RNC (e.g. IMSI).
⇒ The MSC/VLR will use the RANAP: SEC_MODE_CTRL-message to convey setup information related to ciphering and integrity protection towards the RNC. In turn, the RNC will send an RRC: SEC_MODE_CMD-message towards the UE to start the integrity protection procedure and optionally encryption.
⇒ The UE shall confirm start of the integrity protection procedure and of ciphering by sending an RRC: SEC_MODE_COM-message to the RNC.
⇒ The RNC confirms the successful outcome of the RANAP-security mode control procedure by sending the RANAP: SEC_MODE_CTRL-message to the MSC/VLR.

(5) Mobile Originating Conversational Call

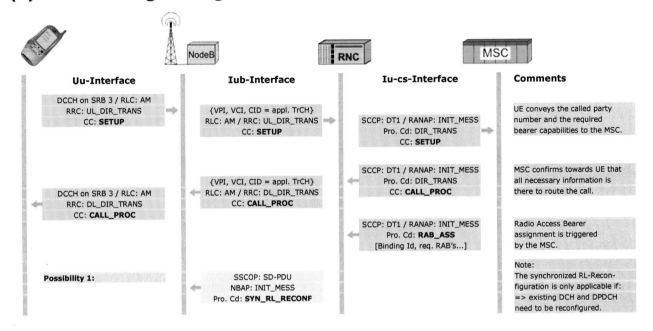

(5) Mobile Originating Conversational Call

Description
⇒ The UE will send a call control (CC) SETUP-message, embedded into an RRC: UL_DIR_TRANS-message, to the MSC, identifying the requested bearer service and the dialed directory number.
⇒ If the MSC has enough information to route the call based on the SETUP-message, it shall send a CC: CALL_PROC-message back to the UE. Note that both the SETUP- and the CALL_PROC-message are transparent for UTRAN.
⇒ Now the MSC will initiate the radio bearer assignment procedures by sending a RANAP: RAB_ASS-message (initiating message) to the SRNC.

From the perspective of the SRNC there are three options to conduct this procedure:

- **Synchronized Radio Link Reconfiguration**
  The synchronized radio link reconfiguration is necessary, if the UTRAN needs to re-define any DCH's and DPCH's. Most importantly, synchronization is required, if the currently used TFCI-values shall be re-used in the new configuration but with a different meaning. This procedure is performed by using the NBAP: SYN_RL_RECONF-messages (initiating message / successful outcome). Since a new radio link is also required for the traffic, this procedure also includes the NBAP: RL_SETUP-messages (initiating message / successful outcome).

- **Unsynchronized Radio Link Reconfiguration**
  The unsynchronized radio link reconfiguration shall be used, if there are already DCH's equipped which shall be re-defined or expanded and if there are only new TFCI-values defined. That is, the former TFCI-values remain in use and therefore, a fine time synchronization to the new configuration is not required. This procedure is performed by using the NBAP: UNSYN_RL_RECONF-messages (initiating message / successful outcome). Since a new radio link is also required for the traffic, this procedure also includes the NBAP: RL_SETUP-messages (initiating message / successful outcome).

- **No Radio Link Reconfiguration**
  This option is applicable, if the UE was using common TrCH's previously. In this case, there is no reconfiguration of the existing radio links but a complete new radio setup. Therefore, this procedure only consists of the NBAP: RL_SETUP-messages (initiating message / successful outcome).

(6) Mobile Originating Conversational Call

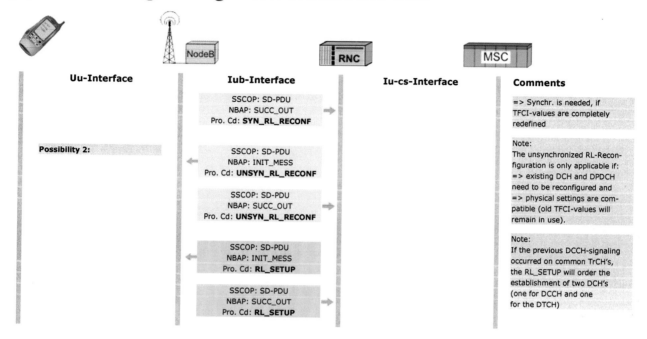

(7) Mobile Originating Conversational Call

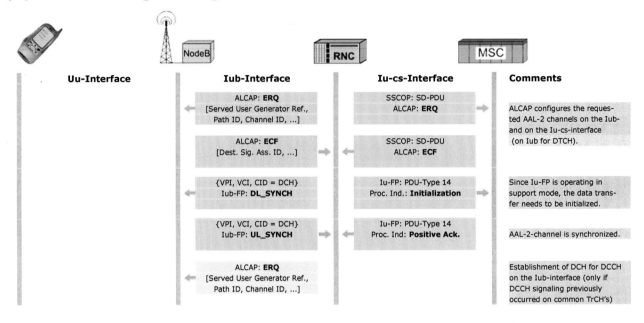

(7) Mobile Originating Conversational Call

Description
⇒ Through ALCAP, the SRNC will establish (additional) AAL-2-links between itself and the MSC and between itself and the NodeB. On Iu-cs-interface, the Iu-FP will perform an initialization of the link. Consecutively, the Iu-FP will be operated in support mode and will exchange (idle) speech frames once every 20 ms.
⇒ On Iub-interface, the Iub-FP also synchronizes the new AAL-2 channel.
⇒ Only if DCCH-signaling prior occurred over common TrCH's, the SRNC will perform a second AAL-2 link establishment for the DCH to carry the DCCH's.

(8) Mobile Originating Conversational Call

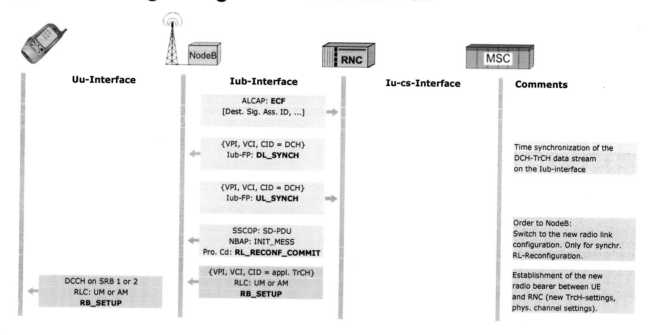

(8) Mobile Originating Conversational Call

Description
⇒ The ALCAP: ECF-message confirms the AAL-2 link establishment.
⇒ On Iub-interface, the Iub-FP also synchronizes the new AAL-2 channel.
⇒ If the synchronized radio link reconfiguration was used, the SRNC shall trigger the switch to the new configuration by sending NBAP: RL_RECONF_COMMIT to the NodeB.
⇒ At the same time, the SRNC will also convey the new radio bearer setup configuration to the UE by sending an RRC: RB_SETUP-message to the UE.

(9) Mobile Originating Conversational Call

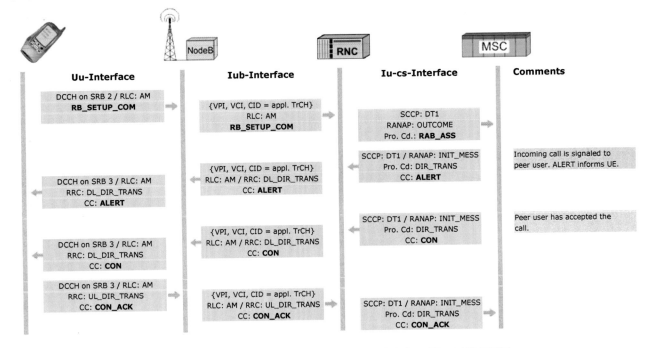

(9) Mobile Originating Conversational Call

Description
⇒ The UE will confirm reception of the RRC: RB_SETUP message and change to the new radio bearer configuration by sending an RRC: RB_SETUP_COM-message back to the SRNC.
⇒ The SRNC finishes the radio bearer assignment procedure by sending RANAP: RAB_ASS-message (outcome) to the MSC/VLR.
⇒ When the called user is alerted of the incoming call, a CC: ALERT-message is sent back to the calling UE. Either this triggers the UE to create a ring-tone locally or the ring-tone is received in-band over the respective TrCH for traffic.
⇒ When the called user answers the phone, a CC: CON-message is sent to the calling UE. The UE shall immediately respond by sending a CC: CON_ACK-message to the MSC/VLR.

(10) Mobile Originating Conversational Call

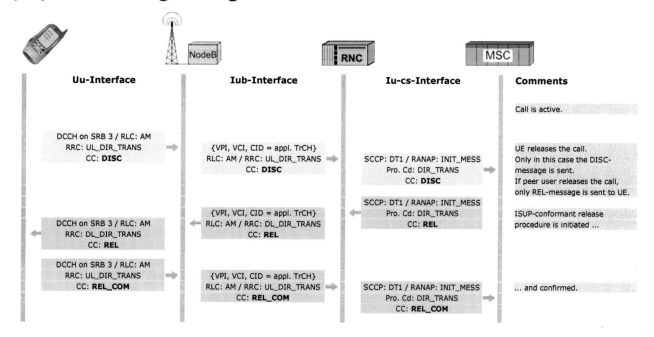

(10) Mobile Originating Conversational Call

Description

> **Note:** The following message is optional. It will be sent,
> - If the UE initiates the call release.

⇒ If the UE releases the call, it will send a CC: DISC-message transparently to the MSC/VLR.

⇒ If the called party released the call and if the MSC/VLR received the CC: DISC-message from the UE, the MSC/VLR shall send a CC: REL-message to the UE.
⇒ The UE shall confirm release of the call control resources by sending a CC: REL_COM-message back to the MSC/VLR.

(11) Mobile Originating Conversational Call

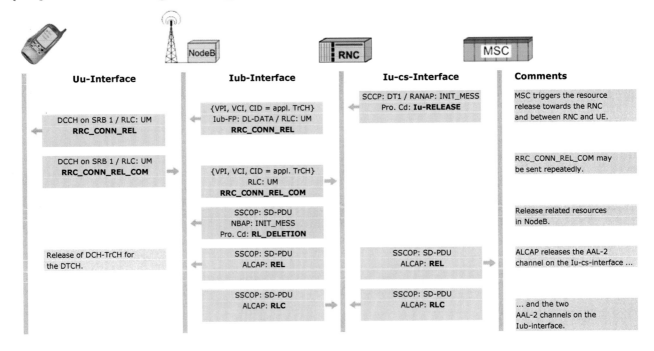

(11) Mobile Originating Conversational Call

Description

⇒ The MSC/VLR will initiate the release of the resources on the Iu-cs-interface by sending a RANAP: Iu-RELEASE-message to the SRNC.
⇒ As we stated under "Initial Conditions" prior in this scenario, the SRNC shall not only release the radio bearers but it will release the entire RRC-connection. This procedure is triggered by sending the RRC: RRC_CONN_REL-message to the UE.
⇒ Consequently, the RNC will send one or more RRC_CONN_REL-message to the UE on SRB 1 (RLC-UM shall be used, since the UE is in CELL_DCH-state).
⇒ The UE confirms release of he RRC-connection and state change back to RRC-idle by sending an RRC_CONN_REL_COM-message to the RNC. As the comment indicates, the RRC_CONN_REL_COM-message will be sent using SRB 1 (⇔ RLC-UMD-PDU's). Since the UE is in CELL_DCH-state, the UE shall retransmit the RRC_CONN_REL_COM-message up to N308 times until the physical and transport channel resources are released.
⇒ The RNC will send an RL_DELETION-message to the NodeB to initiate the release of the dedicated TrCH and of the physical resources between NodeB and UE.
⇒ Through the ALCAP-protocol (REL-messages), the RNC will initiate the release of the related AAL-2 channels on the Iub-interface and on the Iu-cs-interface.
⇒ The ALCAP-protocol within the NodeB and the MSC/VLR will confirm the release of the AAL-2 channels by sending RLC-messages back to the RNC.

(12) Mobile Originating Conversational Call

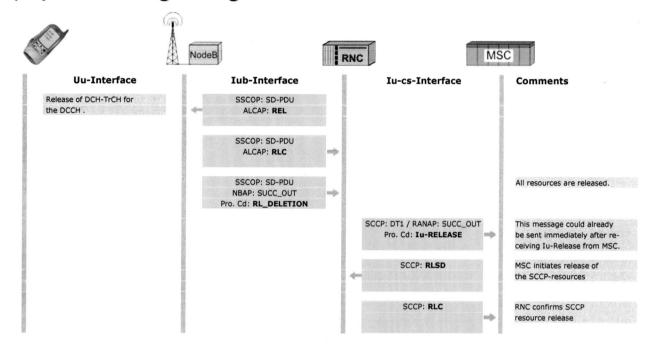

(12) Mobile Originating Conversational Call

Description
⇒ The second ALCAP-message exchange (REL / RLC) is required, because two separate AAL-2 links were established on the Iub-interface for traffic and signaling.
⇒ The NodeB will confirm the release of the dedicated TrCH's and of the physical resources by sending an RL_DELETION-message (successful outcome) back the RNC.
⇒ Latest at this time, the SRNC confirms successful release of the resources on the Iu-cs-interface by sending a RANAP: Iu-RELEASE-message (successful outcome) to the MSC/VLR.
⇒ Finally, the SCCP-connection is released by the MSC by sending an RLSD-message to the RNC. The RNC-confirms by sending an RLC-message back to the MSC. The SCCP-connection release will most likely occur already after the reception of the Iu-RELEASE-message (successful outcome) by the MSC.

(1) Mobile Terminating Conversational Call

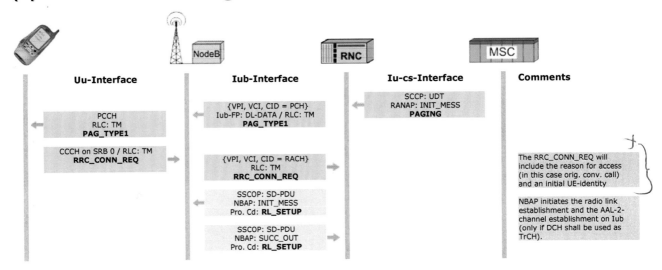

(1) Mobile Terminating Conversational Call

Initial Conditions
The UE is in RRC-idle mode and shall return to RRC-idle mode after successful completion of the procedure. The UE receives a PAG_TYPE1-message from the RNC which triggers the UE to initiate the RRC-connection establishment.

Applicability of this Procedure
This procedure is applicable in case mobile terminating call establishment towards the circuit-switched core network domain when the UE is in RRC-idle mode. The procedure would be very similar, if the UE was in RRC-connected state with the following exceptions:
a) The paging message would be PAG_TYPE2.
b) The UE would not be required to establish an RRC-connection prior to sending the PAG_RSP-message to the core network.
Even if the UE was in RRC-connected state at the time of paging, the establishment of an SCCP-connection would be required to the circuit-switched core network domain (see remark later in this section).

Description
⇒ The UE will start the procedure by sending RRC_CONN_REQ to the RNC.

> **Note:** The following procedure is optional. It will be conducted,
> • If the RNC wants to perform the procedure using a DCH-TrCH in uplink and downlink direction.

⇒ The RNC will send an RL_SETUP-message to the NodeB to request for the establishment of an AAL-2 channel for the DCH-TrCH on the Iub-interface. The NodeB will configure the requested physical resources in uplink and downlink direction and send a NBAP: RL_SETUP-message (successful outcome) back to the CRNC. This message also contains the Binding Id., to be used by the CRNC as "Served User Generator Reference" to link the ALCAP: ERQ-message to the respective RL_SETUP-message.

(2) Mobile Terminating Conversational Call

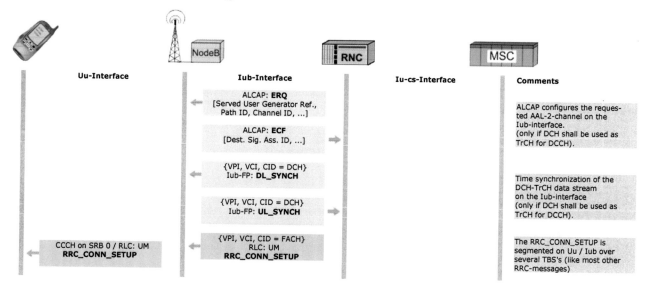

(2) Mobile Terminating Conversational Call

Description

⇒ After positive confirmation of the NodeB, the RNC will send an ALCAP: ERQ-message to the NodeB to physically allocate the AAL-2 channel.
⇒ The NodeB confirms the AAL-2 channel establishment by replying with an ALCAP: ECF-message.
⇒ The RNC will start the TrCH-synchronization by sending an Iub-FP: DL_SYNCH-message to the NodeB. Synchronization is achieved as soon as the NodeB responds with an Iub-FP: UL_SYNCH-message. Finally, the DCH-TrCH is configured. Note that the presented messages are only there, if a DCH-TrCH shall be used for the registration process.

⇒ On the applicable TrCH (FACH), the RNC will convey an RRC_CONN_SETUP-message to the UE. This message will, among other things, configure the signaling radio bearers on DCCH 1 – N plus it will tell the UE whether to move into RRC / CELL_FACH-state (⇔ if common TrCH's shall be used) or into RRC / CELL_DCH-state (⇔ if dedicated TrCH's shall be used).

(3) Mobile Terminating Conversational Call

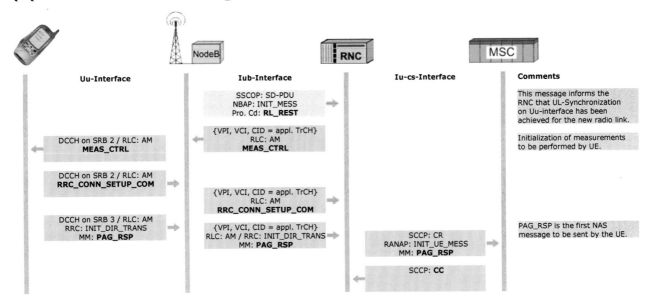

(3) Mobile Terminating Conversational Call

Description

⇒ If DCH's are used, the NodeB will now achieve physical channel synchronization on the new DPCH with the UE. This is indicated towards the RNC by transmitting an RL_REST-message.

⇒ The SRNC will transmit a MEAS_CTRL-message to the UE indicating to the UE which measurements shall be performed.

⇒ If common TrCH's are used, the UE will confirm the RRC-connection establishment by transmitting an RRC_CONN_SETUP_COM-message on CPCH- or RACH-TrCH towards the RNC. Otherwise the uplink DCH-TrCH is used.

⇒ The UE will use the DCCH on SRB 3 to transmit the PAG_RSP-message towards the RNC, embedded in an RRC: INIT_DIR_TRANS-message. The RNC will forward the PAG_RSP-message towards the MSC/VLR.

⇒ Note that an SCCP-connection is required to perform the registration procedure. Besides, RANAP requires an INIT_UE_MESS to establish an Iu signaling connection between the MSC/VLR and the RNC. Accordingly, the PAG_RSP -message will be embedded into a RANAP: INIT_UE_MESS which in turn will be embedded into an SCCP: CR-message. The MSC/VLR will confirm SCCP-connection establishment by sending an SCCP: CC-message to the RNC. Optionally, the SCCP: CR- and CC-messages for connection establishment may be exchanged prior to transmitting the INIT_UE_MESS [PAG_RSP]-message in an SCCP: DT1-message.

Note:
- If the UE was in CELL_DCH-state or CELL_FACH-state at the beginning of the call establishment procedure, the entire RRC-connection establishment would not have been necessary. However, the UE would still need to send the PAG_RSP-message as indicated.
- If the UE was in CELL_PCH-state or URA_PCH-state at the beginning of the call establishment procedure, the UE would initially perform a cell update scenario (cause: paging response), thereby move to CELL_FACH-state and consequently send a PAG_RSP-message to the MSC/VLR as illustrated here.

(4) Mobile Terminating Conversational Call

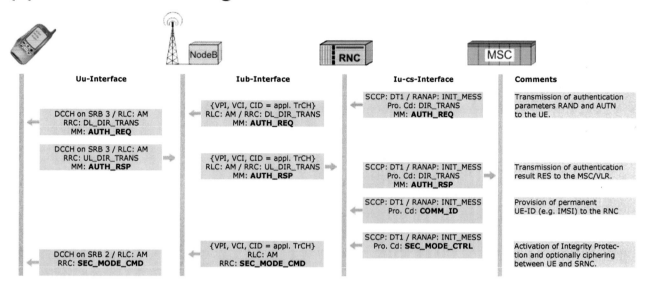

(4) Mobile Terminating Conversational Call

Description
⇒ The MSC/VLR will initiate the user authentication by sending an AUTH_REQ-message to the UE.
⇒ The UE will calculate the authentication result RES and transmit it back the MSC/VLR.
⇒ At this time, the MSC will provide the permanent NAS (Non-Access-Stratum) UE-identity to the RNC (e.g. IMSI).
⇒ The MSC/VLR will use the RANAP: SEC_MODE_CTRL-message to convey setup information related to ciphering and integrity protection towards the RNC. In turn, the RNC will send an RRC: SEC_MODE_CMD-message towards the UE to start the integrity protection procedure and optionally encryption.

(5) Mobile Terminating Conversational Call

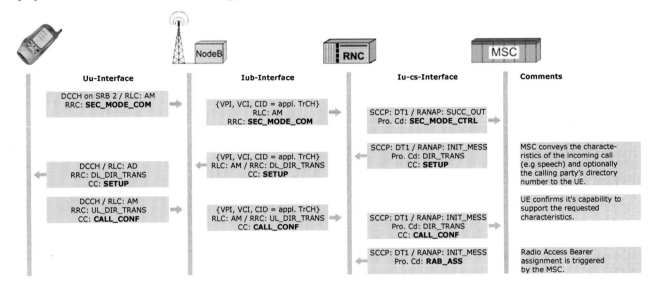

(5) Mobile Terminating Conversational Call

Description
⇒ The UE shall confirm start of the integrity protection procedure and of ciphering by sending an RRC: SEC_MODE_COM-message to the RNC.
⇒ The RNC confirms the successful outcome of the RANAP-security mode control procedure by sending the RANAP: SEC_MODE_CTRL-message to the MSC/VLR.
⇒ The MSC will send a call control (CC) SETUP-message, embedded into an RRC: DL_DIR_TRANS-message, to the UE, identifying the requested bearer service and (if supplementary service CLIP is configured) the dialed directory number.
⇒ If the UE is capable to support the requested service, it shall send a CC: CALL_CONF-message back to the MSC. Note that both the SETUP- and the CALL_CONF-message are transparent for UTRAN.
⇒ Now the MSC will initiate the radio bearer assignment procedures by sending a RANAP: RAB_ASS-message (initiating message) to the SRNC.

(6) Mobile Terminating Conversational Call

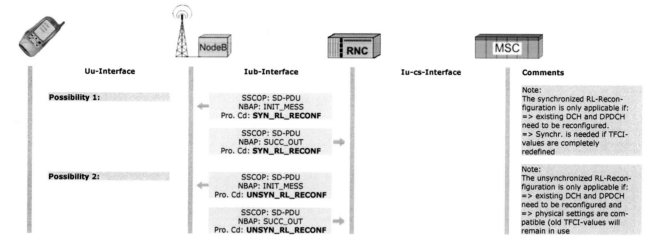

(6) Mobile Terminating Conversational Call

Description
From the perspective of the SRNC there are three options to conduct radio bearer assignment procedure:

- **Synchronized Radio Link Reconfiguration**
  The synchronized radio link reconfiguration is necessary, if the UTRAN needs to re-define any DCH's and DPCH's. Most importantly, synchronization is required, if the currently used TFCI-values shall be re-used in the new configuration but with a different meaning. This procedure is performed by using the NBAP: SYN_RL_RECONF-messages (initiating message / successful outcome). Since a new radio link is also required for the traffic, this procedure also includes the NBAP: RL_SETUP-messages (initiating message / successful outcome).

- **Unsynchronized Radio Link Reconfiguration**
  The unsynchronized radio link reconfiguration shall be used, if there are already DCH's equipped which shall be re-defined or expanded and if there are only new TFCI-values defined. That is, the former TFCI-values remain in use and therefore, a fine time synchronization to the new configuration is not required. This procedure is performed by using the NBAP: UNSYN_RL_RECONF-messages (initiating message / successful outcome). Since a new radio link is also required for the traffic, this procedure also includes the NBAP: RL_SETUP-messages (initiating message / successful outcome).

- **No Radio Link Reconfiguration**
  This option is applicable, if the UE was using common TrCH's previously. In this case, there is no reconfiguration of the existing radio links but a complete new radio setup. Therefore, this procedure only consists of the NBAP: RL_SETUP-messages (initiating message / successful outcome).

(7) Mobile Terminating Conversational Call

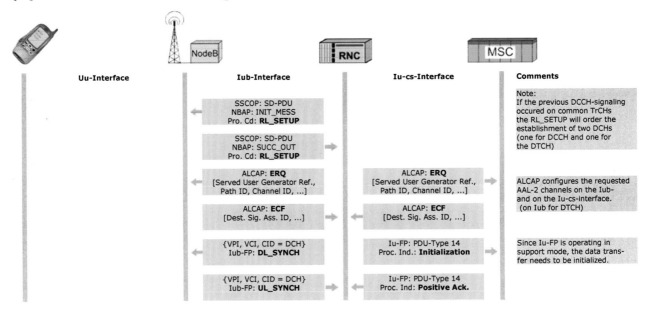

(7) Mobile Terminating Conversational Call

Description
⇒ In either case (synchronized or unsynchronized radio link reconfiguration or previous signaling occurred on common TrCH's) the SRNC needs to send an NBAP: RL_SETUP-message (initiating message) to the NodeB to configure the physical and transport channels.
⇒ Through ALCAP, the SRNC will establish (additional) AAL-2-links between itself and the MSC and between itself and the NodeB. On Iu-cs-interface, the Iu-FP will perform an initialization of the link. Consecutively, the Iu-FP will be operated in support mode and will exchange (idle) speech frames once every 20 ms.
⇒ On Iub-interface, the Iub-FP also synchronizes the new AAL-2 channel.

(8) Mobile Terminating Conversational Call

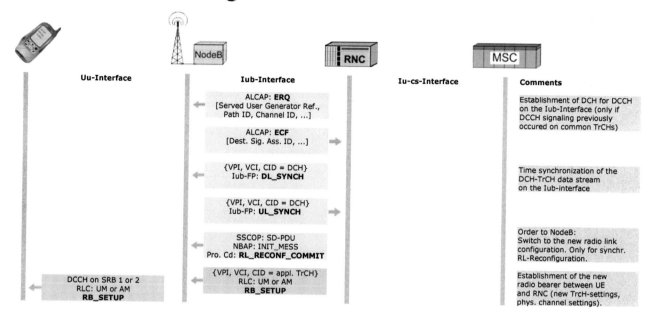

(8) Mobile Terminating Conversational Call

Description
⇒ Only if DCCH-signaling prior occurred over common TrCH's, the SRNC will perform a second AAL-2 link establishment for the DCH to carry the DCCH's.
⇒ If the synchronized radio link reconfiguration was used, the SRNC shall trigger the switch to the new configuration by sending NBAP: RL_RECONF_COMMIT to the NodeB.
⇒ At the same time, the SRNC will also convey the new radio bearer setup configuration to the UE by sending an RRC: RB_SETUP-message to the UE.

(9) Mobile Terminating Conversational Call

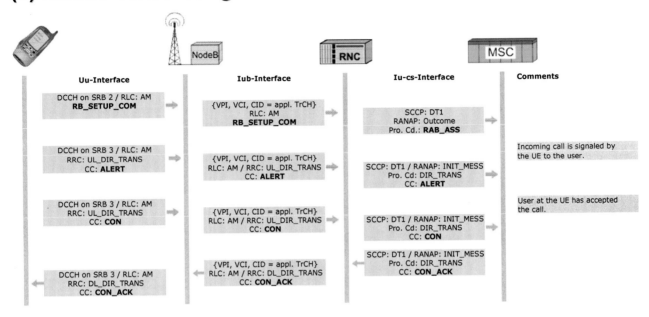

(9) Mobile Terminating Conversational Call

Description
⇒ The UE will confirm reception of the RRC: RB_SETUP message and change to the new radio bearer configuration by sending an RRC: RB_SETUP_COM-message back to the SRNC.
⇒ The SRNC finishes the radio bearer assignment procedure by sending RANAP: RAB_ASS-message (outcome) to the MSC/VLR.
⇒ When the UE alerts the user of the incoming call, a CC: ALERT-message is sent to the MSC.
⇒ When the UE user answers the phone, a CC: CON-message is sent to the MSC. The MSC shall immediately respond by sending a CC: CON_ACK-message to the UE.

Now the call is active.

(10) Mobile Terminating Conversational Call

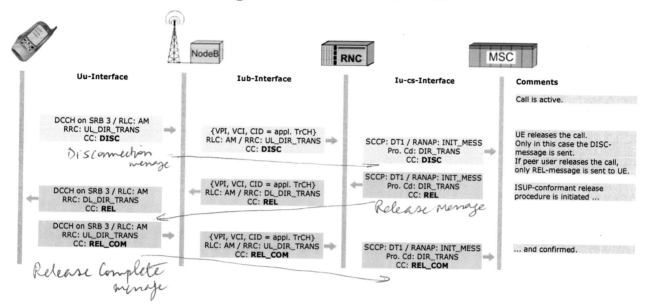

(10) Mobile Terminating Conversational Call

Description

Note: The following message is optional. It will be sent,
- If the UE initiates the call release.

⇒ If the UE releases the call, it will send a CC: DISC-message transparently to the MSC/VLR.

⇒ If the called party released the call and if the MSC/VLR received the CC: DISC-message from the UE, the MSC/VLR shall send a CC: REL-message to the UE.
⇒ The UE shall confirm release of the call control resources by sending a CC: REL_COM-message back to the MSC/VLR.

(11) Mobile Terminating Conversational Call

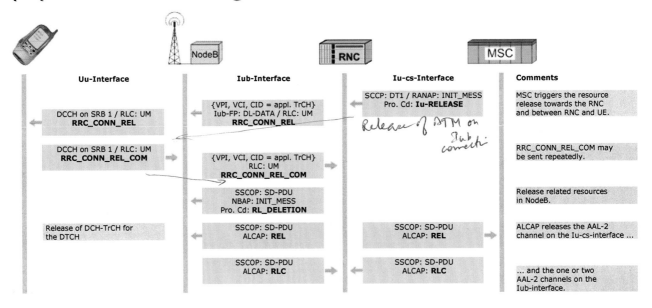

(11) Mobile Terminating Conversational Call

Description
⇒ The MSC/VLR will initiate the release of the resources on the Iu-cs-interface by sending a RANAP: Iu-RELEASE-message to the SRNC.
⇒ As we stated under "Initial Conditions" prior in this scenario, the SRNC shall not only release the radio bearers but it will release the entire RRC-connection. This procedure is triggered by sending the RRC: RRC_CONN_REL-message to the UE.
⇒ Consequently, the RNC will send one or more RRC_CONN_REL-message to the UE on SRB 1 (RLC-UM shall be used, since the UE is in CELL_DCH-state).
⇒ The UE confirms release of the RRC-connection and state change back to RRC-idle by sending an RRC_CONN_REL_COM-message to the RNC. As the comment indicates, the RRC_CONN_REL_COM-message will be sent using SRB 1 (⇔ RLC-UMD-PDU's). Since the UE is in CELL_DCH-state, the UE shall retransmit the RRC_CONN_REL_COM-message up to N308 times until the physical and transport channel resources are released.
⇒ The RNC will send an RL_DELETION-message to the NodeB to initiate the release of the dedicated TrCH and of the physical resources between NodeB and UE.
⇒ Through the ALCAP-protocol (REL-messages), the RNC will initiate the release of the related AAL-2 channels on the Iub-interface and on the Iu-cs-interface.
⇒ The ALCAP-protocol within the NodeB and the MSC/VLR will confirm the release of the AAL-2 channels by sending RLC-messages back to the RNC.

(12) Mobile Terminating Conversational Call

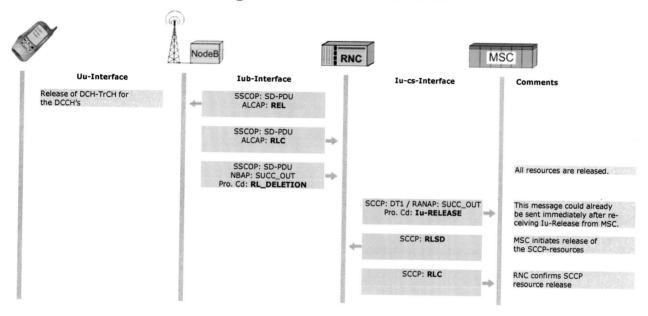

(12) Mobile Terminating Conversational Call

Description
⇒ The second ALCAP-message exchange (REL / RLC) is required, because two separate AAL-2 links were established on the Iub-interface for traffic and signaling.
⇒ The NodeB will confirm the release of the dedicated TrCH's and of the physical resources by sending an RL_DELETION-message (successful outcome) back the RNC.
⇒ Latest at this time, the SRNC confirms successful release of the resources on the Iu-cs-interface by sending a RANAP: Iu-RELEASE-message (successful outcome) to the MSC/VLR.
⇒ Finally, the SCCP-connection is released by the MSC by sending an RLSD-message to the RNC. The RNC-confirms by sending an RLC-message back to the MSC. The SCCP-connection release will most likely occur already after the reception of the Iu-RELEASE-message (successful outcome) by the MSC.

Handover Scenarios

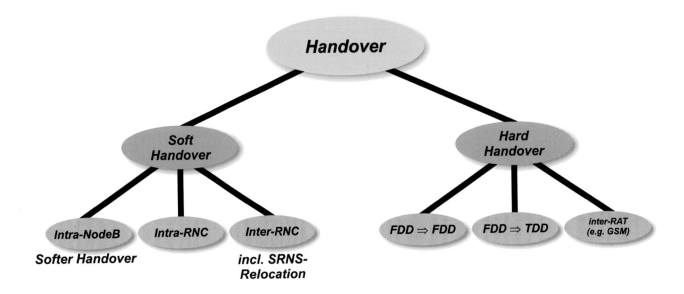

Handover Scenarios

- **Soft Handover Scenarios**
  In a soft handover scenario the UE is not interrupting the link to its serving cell while another radio link is added from a different cell. Soft handover is also applicable when an active radio link is deleted between the UE and one of the serving cells.

  > Note: All soft handover scenarios are intra-frequency handover scenarios.

  The following soft handover scenarios have to be distinguished:
  ⇒ Softer Handover / Intra NodeB
  ⇒ Soft Handover / Intra-RNC and Inter-NodeB
  ⇒ Soft Handover / Inter RNC (⇔ incl. SRNS-Relocation)

- **Hard Handover Scenarios**
  Hard handover scenarios are all handover scenarios in which the active radio link is replaced by another radio link, most likely in a different cell and possibly in a different RAT (radio access technology). The following hard handover scenarios have to be distinguished:
  ⇒ Handover from FDD-Frequency 1 to FDD Frequency 2
  ⇒ Handover from FDD to TDD
  ⇒ Handover from UTRAN to another RAT (e.g. GSM)

  > **Note:**
  > - All soft and hard handover scenarios are only applicable in CELL_DCH-state and are always controlled by the SRNC.
  > - In CELL_DCH-state, the SRNC may also order a cell update procedure to GSM/GPRS by sending an RRC: CELL_CHAN_UTRAN-message to the UE, if there are no DCH's configured towards the circuit-switched core network domain.
  > - In CELL_FACH-state, both the SRNC and the UE may decide for a cell update procedure. In this case, UTRAN would send an RRC: CELL_CHAN_UTRAN-message to the UE or, if the UE decides autonomously, the UE will just perform a cell reselection to the respective GSM-cell and perform a GPRS-cell update procedure or a routing area update procedure, if the GSM-cell belongs to a different routing area.

Variations of Soft Handover

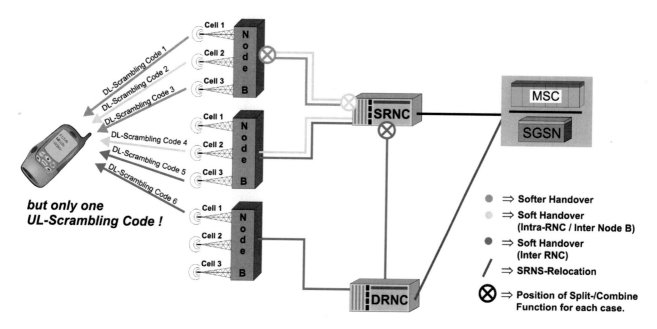

Variations of Soft Handover

For soft handover, different variations exist. Depending on the variation, the split/combine function in uplink direction (network side) will be located either in the NodeB or in the SRNC.

Note:
- In a soft handover situation, the UE shall be able to support a minimum of 6 active radio links to 6 different cells simultaneously [3GTS 25.133 (5.1.2.1)].
- Each cell will use another scrambling code in downlink direction while the UE will use only one uplink scrambling code.
- At least in theory, each of these cells may be connected to another RNC.
- With SSDT (Site Selective Diversity Transmission) there can be maximum 8 simultaneously received cells [3GTS 25.214 (5.2.1.4)].

SRNS-relocation is only indirectly related to soft handover: An SRNS-relocation will be initiated by the SRNC when a (hard or soft) handover procedure results in the radio link removal of the last cell which is controlled by the SRNC.

Note: A DRNC may be connected to a separate MSC/VLR than the SRNC.

(1) Softer Handover (Radio Link Addition)

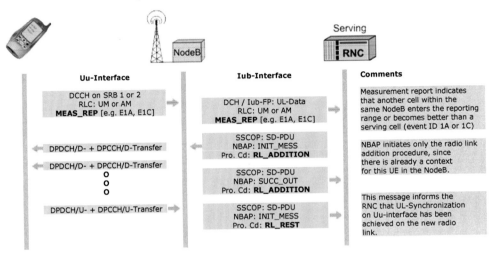

(1) Softer Handover (Radio Link Addition)

Initial Conditions
The UE is in CELL_DCH-state and is connected to one or more NodeB's. The SRNC has previously configured intra-frequency measurements for the UE. Measurement reports shall be sent event triggered. Intra-frequency measurements indicate that another cell enters the reporting range (⇔ Event 1A) or that another non-active cell becomes better than an active cell (⇔ Event 1C). This triggers the transmission of a MEAS_REP-message.

Applicability of this Procedure
The presented procedure is only applicable if the new radio link is established within a NodeB where the UE already has setup one or more active links to cells of this NodeB. Obviously, this remains transparent to the UE.

Description
⇒ The UE will initiate the procedure by sending a MEAS_REP-message to the SRNC. The MEAS_REP-message is sent because another non-active cell enters the reporting range (⇔ Event 1A) or another non-active cell becomes better than an active cell (⇔ Event 1C). The measurements are based on the CPICH RSCP or the CPICH Ec/No of the different cells.
⇒ The SRNC will send an NBAB: RL_ADDITION-message (initiating message) to the NodeB that controls the cell to be added to the active link set. The NBAP: Radio Link Addition procedure is used because there is already a communication context existing between the SRNC and the NodeB for this UE (⇔ softer handover). Accordingly, the NBAP: RL_ADDITION-message will include the same CRNC-Communication Context ID that the previous NBAP-messages from the SRNC to the NodeB for this UE related to.
⇒ Among other things, the RL_ADDITION-message identifies the UE (and its scrambling and channelization code) within the NodeB and it tells the NodeB which scrambling code to use in the target cell in downlink direction (most likely DL-Scrambling Code = 0 which relates to the primary scrambling code of the cell).
⇒ In the case of softer handover, there is no new AAL-2 link nor DCH established between the NodeB and the SRNC because the split/combine-function will reside in the NodeB rather than in the SRNC.
⇒ After the reception of the NBAB: RL_ADDITION-message (initiating message) the NodeB will start to transmit data on DPDCH/D and DPCCH/D to the UE. Note that his data cannot yet be received by the UE, since it is still unaware of the new radio link. The NodeB will also confirm the radio link addition procedure by sending an NBAB: RL_ADDITION-message (successful outcome) to the SRNC.
⇒ When the NodeB has successfully achieved uplink radio link synchronization on the Uu-interface it will send an NBAP: RL_REST-message (initiating message) to the SRNC. Note that in this case synchronization procedure B is used, because there have already been one or more active links prior to the synchronization of this active link.

Note: At this time, radio link synchronization has only been achieved in uplink direction but not in downlink direction.

(2) Softer Handover (Radio Link Addition)

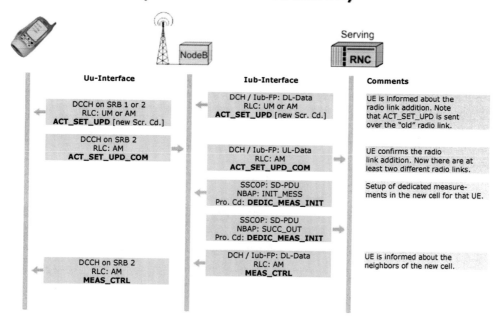

(2) Softer Handover (Radio Link Addition)

Description

⇒ The SRNC will initiate dedicated uplink measurements for that UE in the new cell by sending one or more NBAP: DEDIC_MEAS_INIT-messages (initiating message) to the NodeB which controls the new cell.
⇒ The NodeB will confirm measurement setup by sending NBAP: DEDIC_MEAS_INIT-messages (successful outcome) back to the SRNC.
⇒ Consequently, the SRNC will build an RRC: ACT_SET_UPD-message and send it to the UE using the already established (old) downlink DCH to the UE. Most importantly, this message identifies the scrambling code of the cell which has been added to the active link set.
⇒ The UE will immediately perform physical layer synchronization procedure B to synchronize to the new radio link.
⇒ Without waiting for a successful downlink radio link synchronization of the new radio link the UE shall send an RRC: ACT_SET_UPD_COM-message to the RNC.
⇒ Finally, the SRNC will update the measurement configuration of the UE with respect to the new active cell (new neighbor cells) by sending a MEAS_CTRL-message to the UE.

Softer Handover (Radio Link Removal)

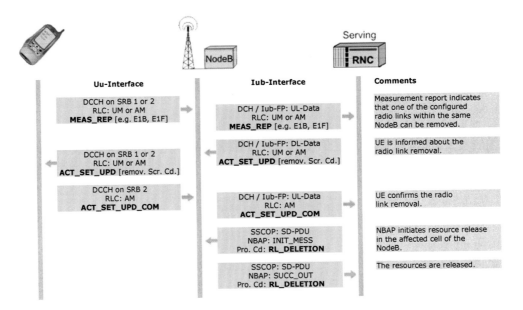

Softer Handover (Radio Link Removal)

Initial Conditions
The UE is in CELL_DCH-state and is connected to one or more NodeB's. The SRNC has previously configured intra-frequency measurements for the UE. Measurement reports shall be sent event triggered. Intra-frequency measurements indicate that the CPICH RSCP or the CPICH Ec/No of one of the monitored cells is leaving the reporting range (⇔ Event 1B) or is underneath a pre-defined threshold (⇔ Event 1F).

Applicability of this Procedure
The presented procedure is only applicable if the radio link to be removed is belonging to a cell which is part of a NodeB to which the UE has still other radio links setup in other cells.

Description
⇒ The UE will initiate the procedure by sending a MEAS_REP-message to the SRNC. The cell for which the UE has indicated Event 1B or 1F has an active link to the UE. This link shall be removed
⇒ Therefore, the SRNC will build an RRC: ACT_SET_UPD-message (radio link removal) and send it to the UE. Most importantly, this message will indicate to the UE which link of the active link set shall be removed. The identification is based on the scrambling code of the link to be removed.
⇒ The UE shall stop receiving data on that radio link and confirm reception of the ACT_SET_UPD-message by sending an RRC: ACT_SET_UPD_COM-message back to the SRNC.
⇒ To release the downlink resources, the SRNC will send an NBAP: RL_DELETION-message (initiating message) to the NodeB that controls the cell to be removed from the active link set. The NBAP: RL_DELETION-message will include the same CRNC- / NodeB-Communication Context ID that the previous NBAP-messages from the SRNC to the NodeB for this UE related to.
⇒ The NodeB will stop to transmit data to the UE on this cell and confirm the radio link deletion procedure by sending an NBAB: RL_DELETION-message (successful outcome) to the SRNC.

Example: ACT_SET_UPD-Message (Link Removal)

```
+---------+------------------------------------------------+--------------------------------------+
|BITMASK  |ID Name                                         |Comment or Value                      |
+---------+------------------------------------------------+--------------------------------------+
|TS 29.331 DCCH-DL (2002-03) (RRC_DCCH_DL)  activeSetUpdate (= activeSetUpdate)                    |
|dL-DCCH-Message                                                                                   |
|1 integrityCheckInfo                                                                              |
|**b32*** |1.1 messageAuthenticationCode                   |                                      |
|-0101--- |1.2 rrc-MessageSequenceNumber                   |1000100101111010100100'B              |
|2 message                                                                                         |
|2.1 activeSetUpdate                                                                               |
|2.1.1 r3                                                                                          |
|2.1.1.1 activeSetUpdate-r3                                                                        |
|***b2*** |2.1.1.1.1 rrc-TransactionIdentifier             |0                                     |
|-1010011 |2.1.1.1.2 maxAllowedUL-TX-Power                 |33                                    |
|2.1.1.1.3 rl-RemovalInformationList                                                               |
|2.1.1.1.3.1 primaryCPICH-Info                                                                     |
|***b9*** |2.1.1.1.3.1.1 primaryScramblingCode             |39                                    |
|----00-- |2.1.1.1.4 tx-DiversityMode                      |noDiversity                           |
+----+--+--+--+--+--+--+--+--+--+--+--+--+--+--+--+--+                                             
|HEX |0 |1 |2 |3 |4 |5 |6 |7 |8 |9 |A |B |C |D |E |F |
+----+--+--+--+--+--+--+--+--+--+--+--+--+--+--+--+--+
|0   |d1|b1|2f|52|28|00|2c|53|02|70|  |  |  |  |  |  |
+----+--+--+--+--+--+--+--+--+--+--+--+--+--+--+--+--+
```

Callouts:
- ACT_SET_UPD-message is used for radio link addition and radio link removal (in this case for radio link removal).
- Identification of the removed radio link (primary scrambling code of cell to be removed from active link set).

Example: ACT_SET_UPD-Message (Link Removal)

Intentionally left blank

(1) Soft Handover (Intra-RNC / Inter-NodeB / Branch Addition)

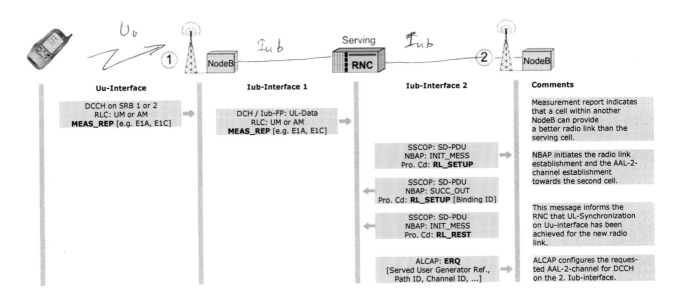

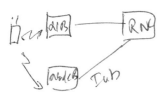

(1) Soft Handover (Intra-RNC / Inter-NodeB / Branch Addition)

Initial Conditions
The UE is in CELL_DCH-state and is connected to one or more NodeB's. The SRNC has previously configured intra-frequency measurements for the UE. Measurement reports shall be sent event triggered. Intra-frequency measurements indicate that another cell enters the reporting range (⇔ Event 1A) or that another non-active cell becomes better than an active cell (⇔ Event 1C). This triggers the transmission of a MEAS_REP-message.

Applicability of this Procedure
The presented procedure is only applicable if the new radio link is established within a NodeB where the UE has not yet setup any radio links to cells of this NodeB. Obviously, this remains transparent to the UE.

Description
⇒ The UE will initiate the procedure by sending a MEAS_REP-message to the SRNC. The MEAS_REP-message is sent because another non-active cell enters the reporting range (⇔ Event 1A) or another non-active cell becomes better than an active cell (⇔ Event 1C). The measurements are based on the CPICH RSCP or the CPICH Ec/No of the different cells.
⇒ The SRNC (which is also the CRNC of the new NodeB) will send an NBAB: RL_SETUP-message (initiating message) to the NodeB that controls the cell to be added to the active link set. The NBAP: Radio Link Setup procedure is necessary because a new communication context needs to be established between the SRNC and the NodeB for this UE (⇔ soft handover).
⇒ Among other things, the RL_SETUP-message identifies the UE (and its scrambling and channelization code) within the NodeB and it tells the NodeB which scrambling code to use in the target cell in downlink direction (most likely DL-Scrambling Code = 0 which relates to the primary scrambling code of the cell).
⇒ After the reception of the NBAB: RL_SETUP-message (initiating message) the NodeB will already start to transmit data on DPDCH/D and DPCCH/D to the UE. Note that his data cannot yet be received by the UE, since it is still unaware of the new radio link. Note: This part is not shown here.
⇒ The NodeB confirms that a new radio link is setup by sending an NBAB: RL_SETUP-message (successful outcome) to the SRNC. Among other things, this message contains the *Binding ID's* for the AAL-2 channels to be established by ALCAP.
⇒ When the NodeB has successfully achieved uplink radio link synchronization on the Uu-interface it will send an NBAP: RL_REST-message (initiating message) to the SRNC. Note that in this case synchronization procedure B is used, because there have already been one or more active links prior to the synchronization of this active link.
⇒ New AAL-2 links for the dedicated TrCH's need to be established between the SRNC and the NodeB. Accordingly, the SRNC will send an ALCAP: ERQ-message to the NodeB to initiate the activation of the first AAL-2-link (in our example for the DCCH but it could also be for the DTCH).

(2) Soft Handover (Intra-RNC / Inter-NodeB / Branch Addition)

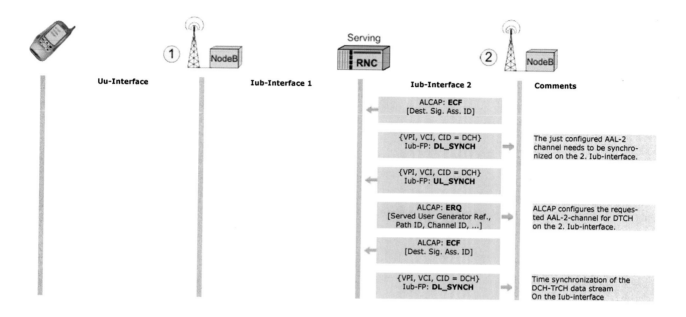

(2) Soft Handover (Intra-RNC / Inter-NodeB / Branch Addition)

Description
⇒ The ALCAP: ECF-message confirms the AAL-2-link establishment for the first DCH.
⇒ Consequently, the new DCH is synchronized on the Iub-interface.
⇒ Using the second *Binding ID* from the NBAP: RL_SETUP-message (successful outcome) as *Served User Generator Reference*, the SRNC will send another ALCAP: ERQ-message to the NodeB to configure another AAL-2-link for the second DCH to be established.
⇒ The ALCAP: ECF-message confirms the AAL-2-link establishment for the second DCH.
⇒ Consequently, the new DCH is synchronized on the Iub-interface.

(3) Soft Handover (Intra-RNC / Inter-NodeB / Branch Addition)

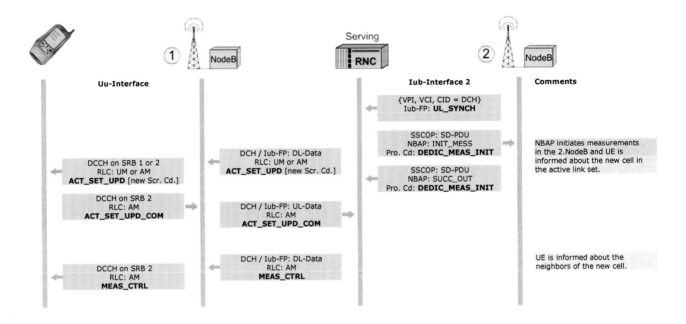

(3) Soft Handover (Intra-RNC / Inter-NodeB / Branch Addition)

Description
⇒ The SRNC will initiate dedicated uplink measurements for that UE in the NodeB which controls the new cell by sending one or more NBAP: DEDIC_MEAS_INIT-messages (initiating message) to that NodeB.
⇒ The NodeB will confirm measurement setup by sending NBAP: DEDIC_MEAS_INIT-messages (successful outcome) back to the SRNC.
⇒ The SRNC will build an RRC: ACT_SET_UPD-message and send it to the UE using the already established (old) downlink DCH to the UE. Most importantly, this message identifies the scrambling code of the cell which has been added to the active link set.
⇒ The UE will immediately perform physical layer synchronization procedure B to synchronize to the new radio link.
⇒ Without waiting for a successful downlink radio link synchronization of the new radio link the UE shall send an RRC: ACT_SET_UPD_COM-message to the RNC.
⇒ Finally, the SRNC will update the measurement configuration of the UE with respect to the new active cell (new neighbor cells) by sending a MEAS_CTRL-message to the UE.

(1) Soft Handover (Inter-RNC / Branch-Addition)

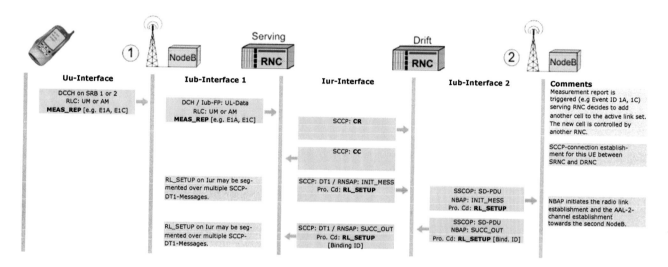

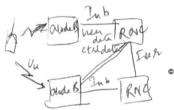

(1) Soft Handover (Inter-RNC / Branch-Addition)

Initial Conditions
The UE is in CELL_DCH-state and is connected to one or more NodeB's. The SRNC has previously configured intra-frequency measurements for the UE. Measurement reports shall be sent event triggered. Intra-frequency measurements indicate that another cell enters the reporting range (⇔ Event 1A) or that another non-active cell becomes better than an active cell (⇔ Event 1C). This triggers the transmission of a MEAS_REP-message.

Applicability of this Procedure
The presented procedure is only applicable:
⇒ if the new radio link is established within a NodeB that is controlled by another RNC than the SRNC.
⇒ if there is no AAL-2-link yet established between that DRNC and the SRNC for that UE. In such a case, the AAL-2-link establishment on the Iur-interface would not be necessary. Instead, we would see a Radio Link Addition procedure on the Iur-interface.

Description
⇒ The UE will initiate the procedure by sending a MEAS_REP-message to the SRNC. The MEAS_REP-message is sent because another non-active cell enters the reporting range (⇔ Event 1A) or another non-active cell becomes better than an active cell (⇔ Event 1C). The measurements are based on the CPICH RSCP or the CPICH Ec/No of the different cells.
⇒ The SRNC determines from its internal database that this cell is controlled by another RNC. Since there is not yet a relation between the SRNC and the DRNC for that UE, the SRNC will first initiate the establishment of an SCCP-connection towards the DRNC by sending an SCCP: CR-message. The SCCP-connection is required as bearer for the RNSAP-signaling. The DRNC shall respond by sending SCCP: CC-message to the SRNC. The CC-message will contain both local references (*Source Local Reference* and *Destination Local Reference*) which are required to identify the upcoming SCCP / RNSAP-message which are related to that UE.
⇒ On the new SCCP-connection, the SRNC will send an RNSAP: RL_SETUP-message (initiating message) to the DRNC. The content of the RNSAP: RL_SETUP-message is very similar to the content of the NBAP: RL_SETUP-message. In that respect, the RNSAP-protocol serves as transparent bearer for the radio link configuration of the UE towards the 2. NodeB.
⇒ Accordingly, the DRNC will relay the radio link information from the RNSAP: RL_SETUP-message into an NBAP: RL_SETUP-message and send it the 2. NodeB.
⇒ The 2. NodeB confirms that a new radio link is setup by sending an NBAB: RL_SETUP-message (successful outcome) to the DRNC. Among other things, this message contains the *Binding ID's* for the AAL-2 channels to be established by ALCAP on the 2. Iub-interface.
⇒ The DRNC relays this information to the SRNC in an RNSAP: RL_SETUP-message (successful outcome) but it also includes *Binding ID's* for the new AAL-2 channels to be established by ALCAP on the Iur-interface

(2) Soft Handover (Inter-RNC / Branch-Addition)

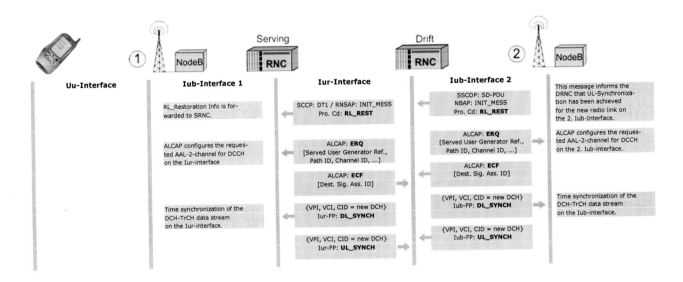

(2) Soft Handover (Inter-RNC / Branch-Addition)

Description
⇒ When the NodeB has successfully achieved uplink radio link synchronization on the Uu-interface it will send an NBAP: RL_REST-message (initiating message) to the DRNC. The DRNC will relay this information through RNSAP to the SRNC.
⇒ New AAL-2 links for the dedicated TrCH's need to be established between the SRNC and the DRNC and between the DRNC and the 2. NodeB.
⇒ ALCAP will take care of this task through the exchange of ALCAP: ERQ- and ECF-messages. Note that the *Binding ID's* from before are used as *Served User Generator References* in the respective ALCAP-messages.
⇒ Consequently, the new DCH's are synchronized on Iub- and Iur-interfaces. One AAL-2 channel is used as bearer of the DCH for the DCCH's while the other AAL-2 channel is used as bearer of the DCH for the DTCH's.

(3) Soft Handover (Inter-RNC / Branch-Addition)

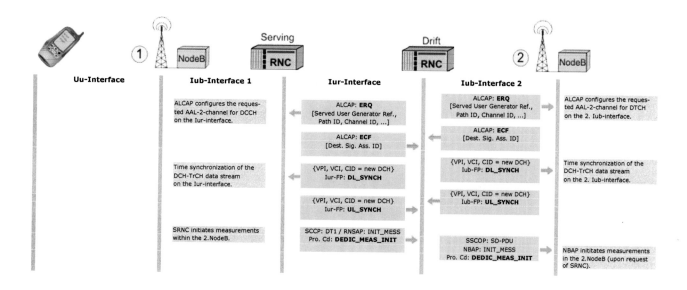

(3) Soft Handover (Inter-RNC / Branch-Addition)

Description
⇒ After the synchronization procedure, the SRNC will convey the dedicated uplink measurement information through RNSAP to the DRNC. The DRNC will relay this information to the 2. NodeB.

(4) Soft Handover (Inter-RNC / Branch-Addition)

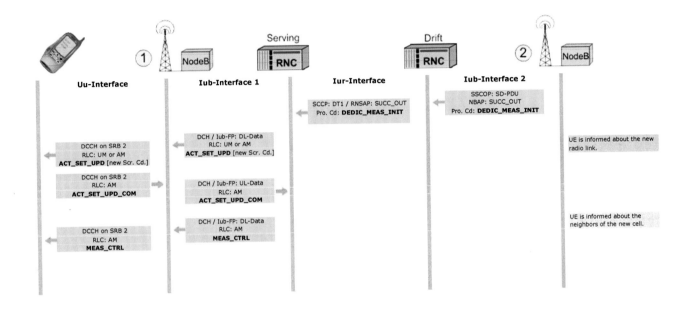

(4) Soft Handover (Inter-RNC / Branch-Addition)

Description

⇒ The 2. NodeB will confirm measurement setup by sending NBAP: DEDIC_MEAS_INIT-messages (successful outcome) back to the DRNC. The DRNC will relay this confirmation back to the SRNC in RNSAP: DEDIC_MEAS_INIT-messages (successful outcome)
⇒ The SRNC will build an RRC: ACT_SET_UPD-message and send it to the UE using the already established (old) downlink DCH to the UE. Most importantly, this message identifies the scrambling code of the cell which has been added to the active link set.
⇒ The UE will immediately perform physical layer synchronization procedure B to synchronize to the new radio link.
⇒ Without waiting for a successful downlink radio link synchronization of the new radio link the UE shall send an RRC: ACT_SET_UPD_COM-message to the RNC.
⇒ Finally, the SRNC will update the measurement configuration of the UE with respect to the new active cell (new neighbor cells) by sending a MEAS_CTRL-message to the UE.

Hard Handover UTRA-FDD ⇒ GSM

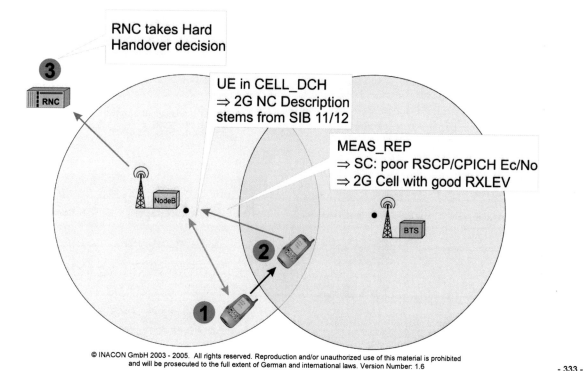

Hard Handover UTRA-FDD ⇒ GSM

Overview
1. The UE is in CELL_DCH-state and moves out of the 3G-coverage. The UE is aware of the adjacent 2G neighbor cell of the 3G serving cell.
2. The UE sends one or more measurement reports through the NodeB to the SRNC which indicate the poor RSCP / CPICH Ec/No compared to a suitable RXLEV for the 2G-neighbor cell. The event ID is either 3A (⇔ UTRAN frequency signal level underneath pre-configured threshold and GSM-frequency signal level is above a pre-configured threshold) or 3C (⇔ the signal level of the GSM-frequency is above a pre-configured threshold).
3. If supported by the UE (UE capabilities) and if supported by the network, the SRNC decides to perform a hard handover towards the GSM neighbor cell.

(1) Hard Handover UTRA-FDD ⇒ GSM (Message Flow)

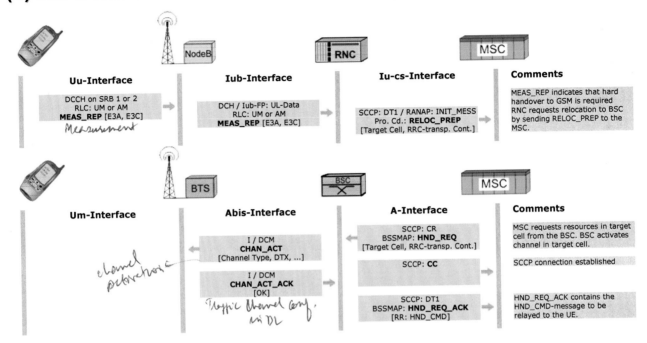

(1) Hard Handover UTRA-FDD ⇒ GSM (Message Flow)

Initial Conditions
The UE is in CELL_DCH state and has a connection only to the circuit-switched core network domain. The UE is about to loose 3G-coverage but there is 2G-coverage (GSM) available.

Applicability of this Procedure
Handover to a GSM-cell is only applicable, if the required radio bearer can be provided by the GERAN and/or if a degradation of the QoS is feasible.

Description
⇒ The UE will indicate through a MEAS_REP-message to the SRNC that 3G-coverage is low but that a 2G-NC could provide a better radio link. The respective event ID is 3A or 3C.
⇒ The SRNC will take a hard handover decision which also requires a relocation of the core network connection to the 2G-BSS. Accordingly, the SRNC will send a RANAP: RELOC_PREP-message to the MSC. This message identifies the target cell and contains an RRC-transparent container for the target BSC which contains, among other things, information about the ciphering method and the currently configured radio bearer mapping. The MSC may have to approach a second MSC, if the target BSS is not connected to this MSC. Inter-MSC-handover scenarios are not covered in this document but in UMTS – Signaling & Protocol Analysis (Volume 2) [8].
⇒ The MSC will send a BSSMAP: HND_REQ-message to the target BSC. This message is embedded into an SCCP: CR-message to establish an SCCP-connection. This CR has to be confirmed by the target BSC through an SCCP: CC-message. The HND_REQ-message contains the transparent container, steaming from the source-RNC, the identification of the target cell and the identification of the A-channel to be used for traffic between BSC and MSC.
⇒ If the BSC is capable to provide the requested bearer channel, it will activate this channel towards the BTS (⇔ CHAN_ACT-message). The BTS shall respond with CHAN_ACT_ACK.
⇒ When all relevant handover data are available (esp. ARFCN, timeslot, Handover Reference), the BSC shall build the RR: HND_CMD-message and embed it into its response message towards the MSC. This response message is the HND_REQ_ACK-message

(2) Hard Handover UTRA-FDD ⇒ GSM (Message Flow)

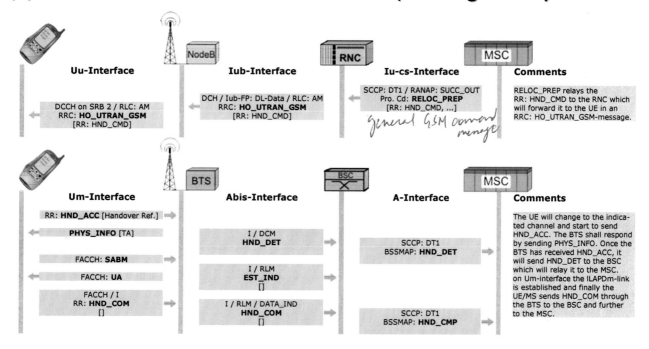

(2) Hard Handover UTRA-FDD ⇒ GSM (Message Flow)

Description

⇒ The MSC will relay the RR: HND_CMD-message into the RANAP: RELOC_PREP-message (successful outcome) and relay it towards the RNC. The RNC will embed the RR: HND_CMD-message into an RRC: HO_UTRAN_GSM-message and send it to the UE.

⇒ The UE will switch to the indicated radio resource (GSM ARFCN & TS, ciphering scheme, ...) and start to transmit access bursts on the allocated ARFCN and TS containing the handover reference (8 bit), which was generated by the target BSS.

⇒ Upon reception of a valid access burst, containing the allocated handover reference, the BTS shall perform two actions:
  1. Transmit a HND_DET-message towards the BSC which will in turn send a BSSMAP: HND_DET-message to the MSC. The reception of this message triggers the switch of the terrestrial resources towards the new radio link.
  2. Transmit PHYS_INFO-messages (generated by the BTS) to the UE in the downlink direction on the allocated TS to convey TA- and power control information to the UE.

⇒ As soon as the UE has received a PHYS_INFO-message it can start using normal bursts and it will transmit a LAPD$_m$: SABM-frame to the BTS to establish the layer 2 link.

⇒ The BTS will confirm layer 2 link establishment by replying a LAPD$_m$: UA-frame to the UE. In addition, the BTS will inform the BSC about the link establishment by sending an EST_IND-message to the BSC.

⇒ Finally, the handover procedure is completed in the target BSS when the UE transmits an RR: HND_COM-message to the BSC which in turn sends a BSSMAP: HND_CMP-message to the MSC.

(3) Hard Handover UTRA-FDD ⇒ GSM (Message Flow)

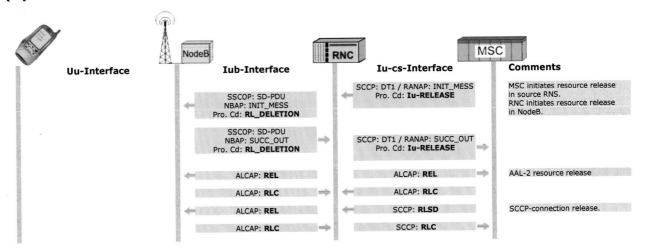

(3) Hard Handover UTRA-FDD ⇒ GSM (Message Flow)

Description
With the reception of the BSSMAP: HND_CMP-message the MSC is aware that the new radio link has successfully been established and that the resources in the source RNS can be released.
Accordingly, the MSC initiates the resource release by sending a RANAP: Iu-RELEASE-message to the RNC.
Consequently, the RNC initiates the resource release on Iub- and Uu-interface by sending an NBAP: RL_DELETION-message to the NodeB. The NodeB releases the resources on the Uu-interface and confirms the resource release by sending an NBAP: RL_DELETION-message (successful outcome) back to the RNC.
Now the RNC will use ALCAP to release the still activated AAL-2 links on Iub-interface and on Iu-cs-interface. At any time after the reception of the RANAP: Iu-RELEASE-message, the RNC can send RANAP: Iu-RELEASE (successful outcome) back the MSC.
Finally, the MSC will release the SCCP-connection by sending an SCCP: RLSD-message to the RNC. The RNC confirms the SCCP-connection release by sending an SCCP: RLC-message back to the MSC.

Example: Hard Handover UTRA FDD ⇒ GSM (Message Flow)

- **Extract of RRC: MEAS_REP-Message**

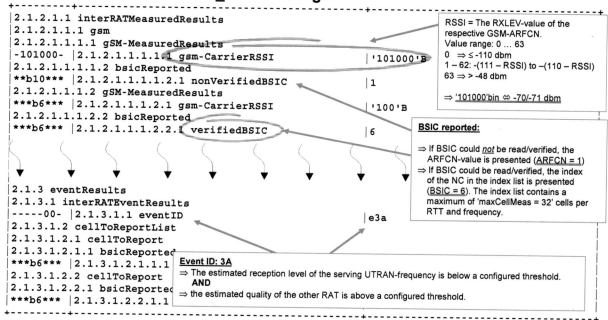

Example: Hard Handover UTRA FDD ⇒ GSM (Message Flow)

- **Extract of RRC: MEAS_REP-Message**
  The figure illustrates an extract of a MEAS_REP-message that is sent by the UE because the reception level of a GSM-frequency (neighbor cell) or GSM-frequencies is above a previously defined threshold and the reception level of the UTRAN-frequency is underneath another predefined reception level threshold.

Example: Hard Handover UTRA FDD ⇒ GSM (Message Flow)

- **Extract of RRC: HO_UTRAN_GSM-Message**

```
+---------+-----------------------------------------+---------------------------------------+
|TS 29.331 DCCH-DL (2002-03) (RRC_DCCH_DL)  handoverFromUTRANCommand-GSM                    |
|handoverFromUTRANCommand-GSM)   dL-DCCH-Message                                            |
|1 integrityCheckInfo                                                                       |
|**b32***  |1.1 messageAuthenticationCode            |'00001101101010000111001011100000'B  |
|-0110---  |1.2 rrc-MessageSequenceNumber            |6                                    |
|2 message                                                                                  |
|2.1 handoverFromUTRANCommand-GSM                                                           |
|2.1.1 r3                                                                                   |
|2.1.1.1 handoverFromUTRANCommand-GSM-r3                                                    |
|------00  |2.1.1.1.1 rrc-TransactionIdentifier      |0                                    |
|2.1.1.1.2 toHandoverRAB-Info                                                               |
|2.1.1.1.2.1 rab-Identity                                                                   |
|***b8***  |2.1.1.1.2.1.1 gsm-MAP-RAB-Identity       |'00000001'B                          |
|--0-----  |2.1.1.1.2.2 cn-DomainIdentity            |cs-domain                            |
|---0----  |2.1.1.1.2.3 re-EstablishmentTimer        |useT314                              |
|----0---  |2.1.1.1.3 frequency-band                 |dcs1800BandUsed                      |
|2.1.1.1.4 gsm-message                                                                      |
|2.1.1.1.4.1 gsm-MessageList                                                                |
|2.1.1.1.4.1.1 gsm-Messages                                                                 |
|**b328**  |2.1.1.1.4.1.1.1 bIT                      |'00000110001010110001100100000010'B  |
|          |                                         |'00011100011000000000010111111000'B  |
```

Embedded RR: HND_CMD-message: The highlighted part contains the Skip Indicator (SI = 0), Protocol Discriminator (PD = 6) and Message Type (MT = 2B) of this message.

Example: Hard Handover UTRA FDD ⇒ GSM (Message Flow)

Intentionally left blank

Hard Handover GSM ⇒ UTRA-FDD

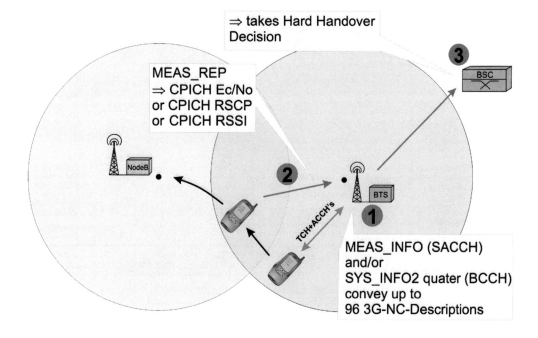

Hard Handover GSM ⇒ UTRA-FDD

Overview
1. The UE has currently activated a GSM-TCH plus the associated control channels (SACCH / FACCH). Through MEAS_INFO-messages and or through SYS_INFO2quater-messages, the BSC informs the UE about the surrounding 3G neighbor cells. Up to 96 different 3G neighbor cells may be identified.
2. The UE is sending normal GSM measurement results towards the BSC which indicate a suitable 3G reception level.
3. If the BSC is equipped to do so, it will initiate a hard handover procedure towards the most suitable UTRA-FDD-cell.

(1) Hard Handover GSM ⇒ UTRA-FDD (Message Flow)

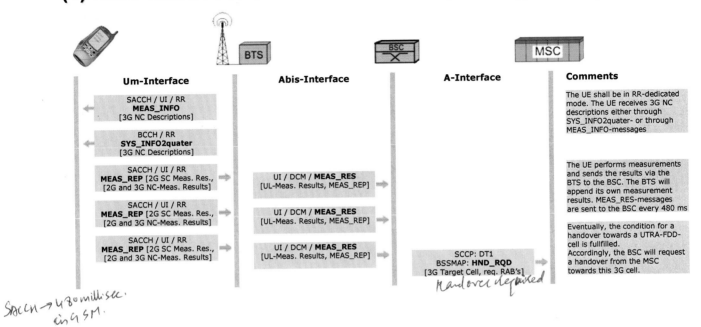

(1) Hard Handover GSM ⇒ UTRA-FDD (Message Flow)

Initial Conditions
The UE has currently a circuit-switched TCH+SACCH+FACCH setup within the GERAN and is connected to a 2G-MSC or to a 3G-MSC. The UE is a multi-mode terminal that supports GSM- and UTRA-FDD-access networks.

Applicability of this Procedure
The presented procedure is applicable in all cases when a UE moves into a 3G-cell coverage area. The BSC is equipped to detect that handover to a 3G-cell will provide a better service. For the purpose of the presented procedure within the BSS it is not relevant whether 1 or 2 MSC's are involved.

Description
⇒ The UE receives 3G-neighbor cell description (UARFCN, Primary Scrambling Code, ...) through the BCCH (⇔ SYS_INFO2quater) and/or through the SACCH (⇔ MEAS_INFO-messages).
⇒ The UE is sending MEAS_REP-messages to the BTS that contain measurement results (⇔ RXLEV, RXQUAL) for the serving 2G-cell and the 2G-NC's as well as measurement results for the 3G-NC's. For the 3G-NC's, the UE shall report either RSCP or CPICH Ec/No, depending on the setting of the parameter FDD_REP_QUANT (0 = RSCP / 1 = CPICH Ec / No). The BTS appends its uplink measurement results and forwards everything to the BSC within MEAS_RES-messages over the Abis-interface. The periodicity of the MEAS_REP/MEAS_RES-message is 480 ms.
⇒ Eventually, the measured value for RSCP or CPICH Ec/No indicates to the BSC that a handover to the most suitable 3G-NC is favorable. At this time, the BSC shall send a HND_RQD-message to the MSC, indicating the very 3G-cell ID to which the handover shall be performed and containing an RRC-transparent container which is destined for the target RNC.

(2) Hard Handover GSM ⇒ UTRA-FDD (Message Flow)

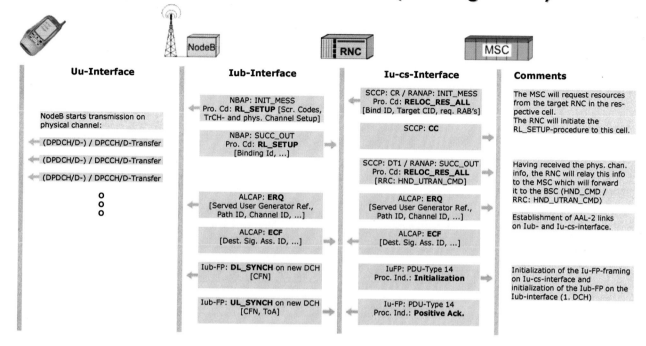

(2) Hard Handover GSM ⇒ UTRA-FDD (Message Flow)

Description

⇒ The MSC possibly has to contact the MSC which controls the respective 3G-cell (⇔ Inter-MSC Handover) or it can directly send a RANAP: RELOC_RES_ALL-message to the RNC which controls the 3G target cell (⇔ Intra-MSC Handover). The Inter-MSC-procedures for handover are part of the volume 2 of UMTS – Signaling & Protocol Analysis [8]. In either case, the MSC will embed the RELOC_RES_ALL-message into an SCCP. CR-message to establish an SCCP-connection on Iu-cs-interface.

⇒ The RELOC_RES_ALL-message will contain, among other things: 1. The Binding ID's for the transport channels (AAL-2) to be established on the Iu-cs-interface. 2. The specification of the radio bearers to be established 3. The identification of the target cell. This information is embedded in an RRC-transparent container.

⇒ The RNC shall confirm the SCCP-connection establishment by sending SCCP: CC to the MSC. In addition, the RNC will initiate the setup of the radio link and the AAL-2 channel on the Iub-interface by sending NBAP: RL_SETUP to the NodeB which contains the target cell.

⇒ The NodeB will configure the requested physical channels on Uu-interface and start transmitting in downlink direction. The NodeB will confirm the setup of the radio link by sending NBAP: RL_SETUP (successful outcome) to the RNC. This message also contains the Binding ID as identification of the transport channel (AAL-2) on the Iub-interface.

⇒ The RNC confirms the setup of the radio link and the establishment of the other resources to the MSC by sending a RANAP: RELOC_RES_ALL-message (successful outcome) to the MSC. This message contains the 25.331/RRC-message: HND_UTRAN_CMD which is only defined as transparent container over GSM-bearers to be forwarded to the UE. Note that the HND_UTRAN_CMD-message is not applicable on UTRAN logical channels.

⇒ The RNC will through ALCAP establish the AAL-2 channels on Iub-interface and Iu-cs-interface. The IE: Served User Generator Reference relates to the respective Binding ID, prior received by the RNC.

⇒ On Iu-cs-interface, the Iu-FP is initialized.

(3) Hard Handover GSM ⇒ UTRA-FDD (Message Flow)

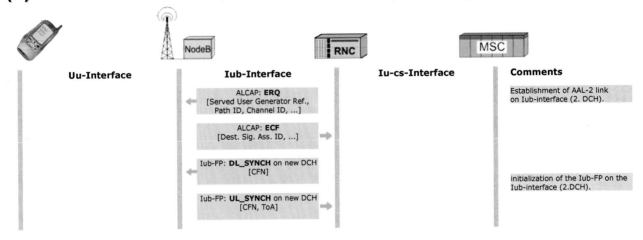

(3) Hard Handover GSM ⇒ UTRA-FDD (Message Flow)

Description
⇒ The second DCH is established for the DCCH-information transfer.

(4) Hard Handover GSM ⇒ UTRA-FDD (Message Flow)

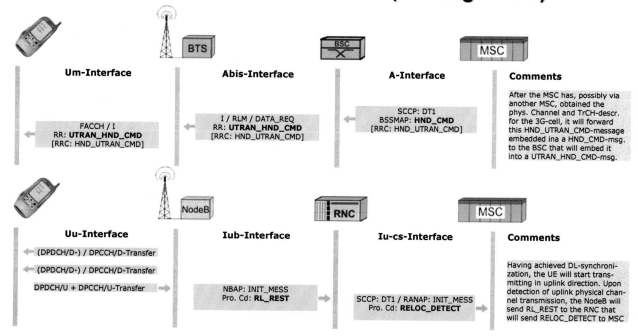

(4) Hard Handover GSM ⇒ UTRA-FDD (Message Flow)

Description

⇒ The MSC shall forward the handover information (RRC: HND_UTRAN_CMD) to the BSC in a BSSMAP: HND_CMD-message.
⇒ The BSC will relay the RRC: HND_UTRAN_CMD-message into an RR: UTRAN_HND_CMD-message and send this message over the Abis-interface and through the BTS to the UE.
⇒ The UE will react on the handover message by switching to the indicated resources. Since there has been no UTRA radio link setup before, the UE shall perform physical synchronization procedure A (⇔ 3GPP 25.214 (4.3.2.3)). The UE will first achieve frame and slot synchronization in downlink direction before starting to transmit in uplink direction.
⇒ Having received valid information on DPDCH/U and DPCCH/U, the NodeB shall confirm physical link establishment towards the RNC by sending the NBAP: RL_REST-message.
⇒ The RNC will confirm link establishment on Uu-interface towards the MSC by sending a RANAP: RELOC_DET-message. Having received the RELOC_DET-message, the MSC will switch the data link to the new radio link.

(5) Hard Handover GSM ⇒ UTRA-FDD (Message Flow)

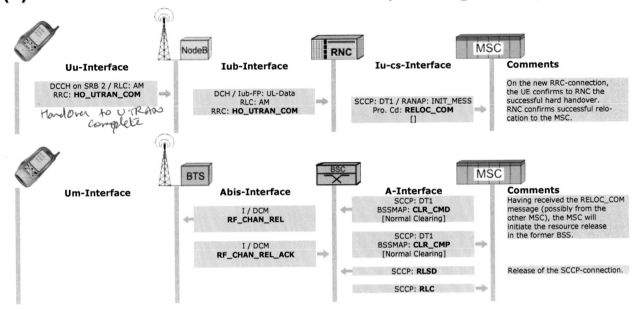

(5) Hard Handover GSM ⇒ UTRA-FDD (Message Flow)

Description
⇒ The UE will use the SRB 2 to transmit an RRC: HO_UTRAN_COM-message to the RNC.
⇒ The RNC confirms successful layer 3-establishment by sending a RANAP: RELOC_COM-message to the MSC.
⇒ The MSC has to trigger the release of the resources in the GSM-BSS by sending a BSSMAP: CLR_CMD-message to the BSC. In turn, the BSC will switch off the radio transmission in the BTS (⇔ RF_CHAN_REL-message) and send a BSSMAP: CLR_CMP-message back to the MSC.
⇒ Finally, the MSC will initiate the SCCP-resource release on the A-interface by sending an SCCP: RLSD-message to the BSC which confirms the SCCP-connection release by replying an SCCP: RLC-message.

UTRAN Mobility Management Procedures

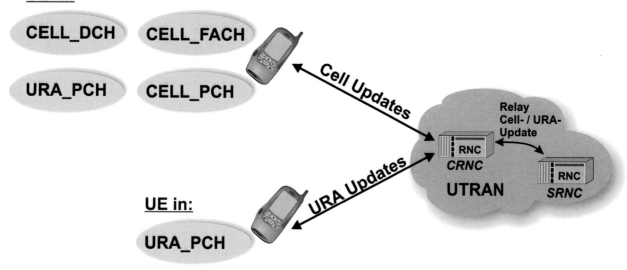

UTRAN Mobility Management Procedures

There are two different UTRAN mobility management procedures:

- **Cell Update Procedures**
  As the figure illustrates, cell update procedures may be conducted in any RRC-connected state. It depends on other conditions (e.g. paging received in URA_PCH-state or CELL_PCH-state, RLC-AM unrecoverable error, new serving cell, ...) whether or not a cell update is actually performed. Another important reason to perform a cell update procedure is periodic cell updating according to T305 which is only applicable in case the UE is in CELL_FACH or CELL_PCH state. More details about when a cell update will be performed can be found on the following pages. However, an exhaustive listing for reasons for cell update scenarios can only be found in [7] and in 3GTS 25.331 (8.3.1).

- **URA Update Procedures**
  URA update procedures are only applicable:
  ⇒ if the UE is in URA_PCH state and selects a new serving cell which belongs to another URA.
  ⇒ If the UE is in URA_PCH state and T305 expires.

> **Note:**
> - Cell update and URA update procedure may concern two RNC's (⇔ CRNC and SRNC) or only one RNC (SRNC).
> - Cell update and URA update procedure may lead to an SRNC-relocation procedure, if the SRNC decides to convey the CN-connection to the second RNC.

[3GTS 25.331 (8.3)]

Cell Update (Intra-RNC / Cell Reselection, Periodic, Page Rsp.)

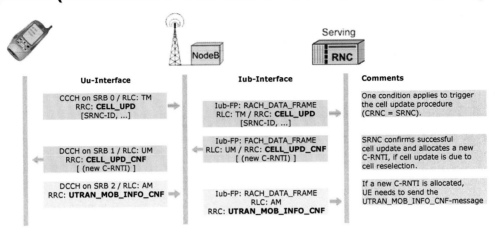

Cell Update (Intra-RNC / Cell Reselection, Periodic, Page Rsp.)

Initial Conditions
The UE is in CELL_FACH, CELL_PCH state or URA_PCH state. One of the following conditions applies:

Applicability of this Procedure
1. The UE selects another serving cell which is connected to the same RNC. Note that the UE is unaware of whether a new serving cell is connected to another RNC. The CRNC which receives the CELL_UPD-message on CCCH can identify the SRNC of the UE through the U-RNTI which is included in the CELL_UPD-message.
2. Timer T305 expires (⇔ periodical cell update).
3. The UE is in CELL_PCH or URA_PCH state and the UE has to transmit uplink data and therefore has to move to CELL_FACH-state.
4. The UE is in CELL_PCH or URA_PCH state, receives a PAG_TYPE1-message and has to send a paging response.
5. The UE has previously lost UTRAN-coverage and now re-enters into a UTRAN service area before timers T307 or T317 expire.
6. The SRNC has previously ordered the UE to move into CELL_FACH state but the UE has no C-RNTI allocated.

Description
⇒ The UE will start the procedure by sending a CELL_UPD-message on CCCH through the NodeB to the SRNC.
⇒ The SRNC will allocate a new C-RNTI, if the cell update is due to cell reselection and will send a CELL_UPD_CNF-message to the UE. This message is by default sent on DCCH but can also be sent on CCCH, if ciphering shall not be performed. If the cell update scenario is due to paging response and if the SRNC needs to move the UE into CELL_DCH state, the CELL_UPD_CNF-message will also allocate and identify the new DCH-TrCH's.
⇒ If the SRNC has allocated a new C-RNTI (only in case of cell update cause: cell reselection), the UE shall respond by sending a UTRAN_MOB_INFO_CNF-message to the SRNC.

(1) Cell Update (Inter-RNC / Cell Reselection / UE initiated)

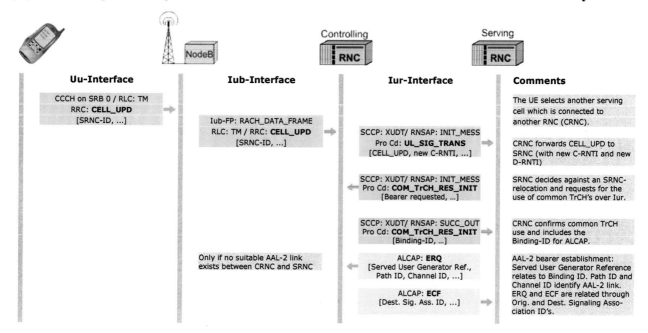

(1) Cell Update (Inter-RNC / Cell Reselection / UE initiated)

Initial Conditions
The UE is in CELL_FACH, CELL_PCH state or URA_PCH state. One of the following conditions applies:

Applicability of this Procedure
1. The UE selects another serving cell which is connected to another RNC Note that the UE is unaware of whether a new serving cell is connected to another RNC. The CRNC which receives the CELL_UPD-message on CCCH can identify the SRNC of the UE through the U-RNTI which is included in the CELL_UPD-message.
2. The UE has previously lost UTRAN-coverage and now re-enters into a UTRAN service area before timers T307 or T317 expire.

In both cases, the SRNC decides not to perform an SRNC-relocation procedure.

Description
⇒ The UE will start the procedure by sending a CELL_UPD-message on CCCH to the CRNC.
⇒ The CRNC will identify the SRNC based on the SRNC-Id which is part of the U-RNTI which is included in the CELL_UPD-message. Accordingly, the RRC-protocol within the CRNC will use the RNSAP: UL_SIG_TRANS-message) to relay the CELL_UPD-message towards the SRNC. The UL_SIG_TRANS-message shall also include the new C-RNTI and D-RNTI which the CRNC allocates for the UE.
⇒ Upon reception of the UL_SIG_TRANS-message, the SRNC becomes aware of the cell reselection of this UE and decides against an SRNC-relocation procedure. To be able to communicate with the UE through common transport channels which are controlled by the CRNC, the SRNC needs to initialize or establish an AAL-2 bearer towards the CRNC. This is done through the RNSAP: COM_TrCH_RES_INIT-message.
⇒ The CRNC confirms the initialization of common transport channel resources to be used and conveys to the SRNC the necessary information to use the required common TrCH's within the CRNC.
⇒ If there is no suitable AAL-2-bearer yet between CRNC and SRNC, the SRNC will request ALCAP to establish this AAL-2-bearer.

(2) Cell Update (Inter-RNC / Cell Reselection / UE initiated)

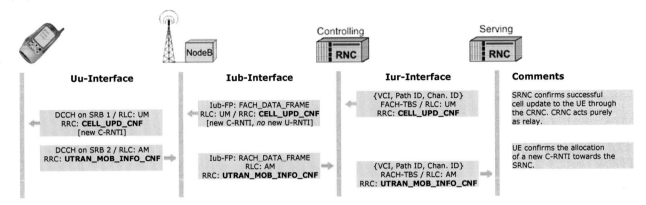

(2) Cell Update (Inter-RNC / Cell Reselection / UE initiated)

Description
⇒ The SRNC will send a CELL_UPD_CNF-message to the UE that contains, among other IE's, the C-RNTI which was allocated by the CRNC. The C-RNTI will allow the UE to use the common TrCH's within the new serving cell.
⇒ Note that the CELL_UPD_CNF-message is sent on the Iur-interface directly as TBS (transport block set) over the AAL-2-bearer (not on Iur-FP). The CRNC will simply relay the CELL_UPD_CNF-message to the UE on the FACH which is related to the RACH that was used by the UE to send the original CELL_UPD-message.
⇒ The CELL_UPD-message is by default sent on DCCH but can also be sent on CCCH, if ciphering shall not be performed.
⇒ Since the CRNC has allocated a new C-RNTI, the UE shall respond by sending a UTRAN_MOB_INFO_CNF-message to the SRNC.

(1) URA Update (Inter-RNC / Cell Reselection / UE initiated)

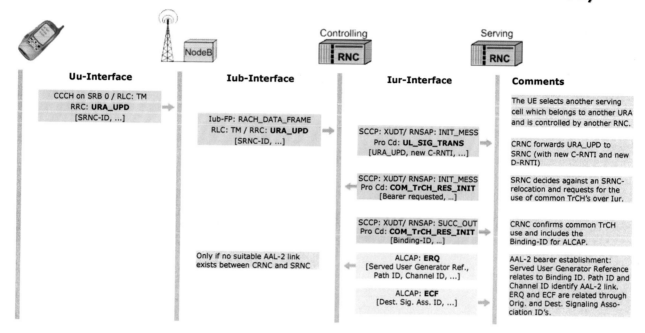

(1) URA Update (Inter-RNC / Cell Reselection / UE initiated)

Initial Conditions
The UE is in URA_PCH state.

Applicability of this Procedure
The UE selects another serving cell which belongs to another URA and which is connected to another RNC. The SRNC decides not to perform an SRNC-relocation procedure.

Description
⇒ The UE will start the procedure by sending a URA_UPD-message on CCCH to the CRNC.
⇒ The CRNC will identify the SRNC based on the SRNC-Id which is part of the U-RNTI which is included in the URA_UPD-message. Accordingly, the RRC-protocol within the CRNC will use the RNSAP: UL_SIG_TRANS-message) to relay the URA_UPD-message towards the SRNC. The UL_SIG_TRANS-message shall also include the new C-RNTI and D-RNTI which the CRNC allocates for the UE.
⇒ Upon reception of the UL_SIG_TRANS-message, the SRNC becomes aware of the cell reselection of this UE and decides against an SRNC-relocation procedure. To be able to communicate with the UE through common transport channels which are controlled by the CRNC, the SRNC needs to initialize or establish an AAL-2 bearer towards the CRNC. This is done through the RNSAP: COM_TrCH_RES_INIT-message.
⇒ The CRNC confirms the initialization of common transport channel resources to be used and conveys to the SRNC the necessary information to use the required common TrCH's within the CRNC.
⇒ If there is no suitable AAL-2-bearer yet between CRNC and SRNC, the SRNC will request ALCAP to establish this AAL-2-bearer.

(2) URA Update (Inter-RNC / Cell Reselection / UE initiated)

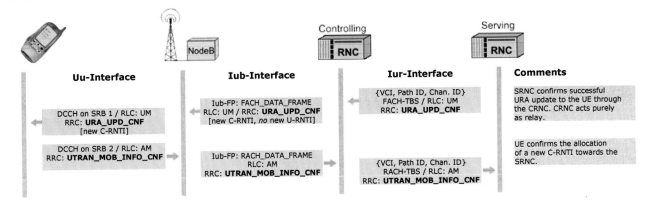

(2) URA Update (Inter-RNC / Cell Reselection / UE initiated)

Description
⇒ The SRNC will send a URA_UPD_CNF-message to the UE that contains, among other IE's, the C-RNTI which was allocated by the CRNC. The C-RNTI will allow the UE to use the common TrCH's within the new serving cell.
⇒ Note that the URA_UPD_CNF-message is sent on the Iur-interface directly as TBS (transport block set) over the AAL-2-bearer (not on Iur-FP). The CRNC will simply relay the URA_UPD_CNF-message to the UE on the FACH which is related to the RACH that was used by the UE to send the original URA_UPD-message.
⇒ The URA_UPD-message is by default sent on DCCH but can also be sent on CCCH, if ciphering shall not be performed.
⇒ Since the CRNC has allocated a new C-RNTI, the UE shall respond by sending a UTRAN_MOB_INFO_CNF-message to the SRNC.

Packet-Switched Scenarios

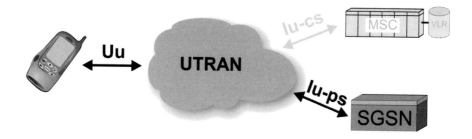

Packet-Switched Scenarios

Packet-switched procedures relate to all procedures that relate to the UE communicating with the packet-switched core network.

(1) Attachment

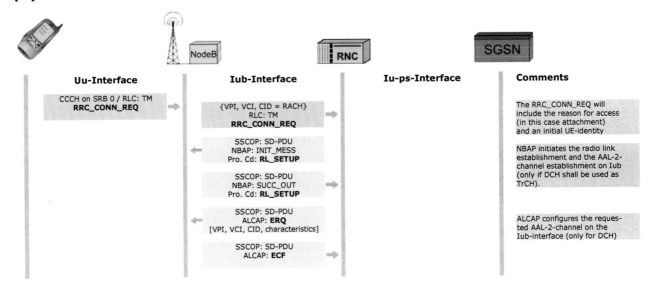

(1) Attachment

Initial Conditions
The UE is initially in RRC-idle mode but shall remain RRC-connected after successful completion of the procedure.
If the network operates in NOM I (⇔ Gs-interface present) then the attachment *may* also include IMSI-attachment. If the network operates in NOM II (⇔ no Gs-interface) then the attachment will be exclusively related to the packet-switched domain.

Applicability of this Procedure
This procedure is applicable in case of packet-switched attachment only or for combined IMSI-attach and packet-switched attach.

Description
⇒ The UE will start the procedure by sending RRC_CONN_REQ to the RNC.

> **Note:** The following procedure is optional. It will be conducted,
> - If the RNC wants to perform the procedures using a DCH-TrCH in uplink and downlink direction.

⇒ The RNC will send an RL_SETUP-message to the NodeB to request for the establishment of an AAL-2 channel on the Iub-interface.
⇒ After positive confirmation of the NodeB, the RNC will send an ALCAP: ERQ-message to the NodeB to physically allocate the AAL-2 channel.
⇒ The NodeB confirms the AAL-2 channel establishment by replying with an ALCAP: ECF-message.

(2) Attachment

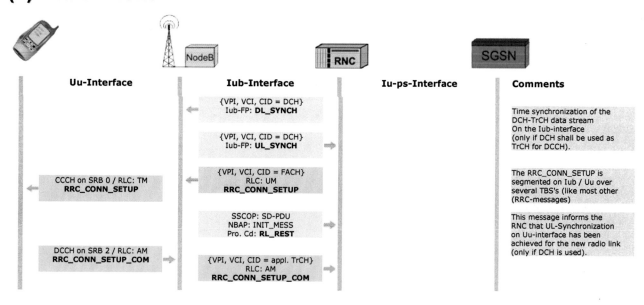

(2) Attachment

Description

⇒ The RNC-will start the TrCH-synchronization by sending an Iub_FP: DL_SYNCH-message to the NodeB. Synchronization is achieved as soon as the NodeB responds with an Iub-FP: UL_SYNCH-message. Finally, the DCH-TrCH is configured. Note that the presented messages are only there, if a DCH-TrCH shall be used for the registration process.

⇒ On the applicable TrCH (FACH), the RNC will convey an RRC_CONN_SETUP-message to the UE. This message will, among other things, configure the signaling radio bearers on DCCH 1 – N plus it will tell the UE whether to move into RRC / CELL_FACH-state (⇔ if common TrCH's shall be used) or into RRC / CELL_DCH-state (⇔ if dedicated TrCH's shall be used).

⇒ If DCH's are used, the NodeB will achieve physical channel synchronization with the UE. This is indicated towards the RNC by transmitting an RL_REST-message.

⇒ If common TrCH's are used, the UE will confirm the RRC-connection establishment by transmitting an RRC_CONN_SETUP_COM-message on CPCH- or RACH-TrCH towards the RNC. Otherwise the uplink DCH-TrCH is used.

(3) Attachment

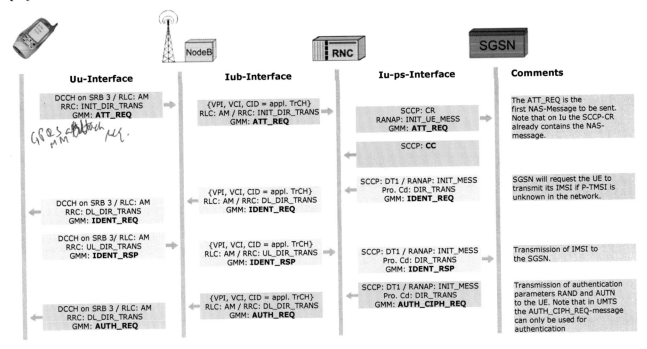

(3) Attachment

Description
⇒ The UE will use the DCCH on SRB 3 to transmit the ATT_REQ-message towards the RNC. The RNC will forward the ATT_REQ-message towards the SGSN.
⇒ Note that an SCCP-connection is required to perform the registration procedure. Besides, RANAP requires an INIT_UE_MESS, to establish an Iu signaling connection between the SGSN and the RNC. Accordingly, the ATT_REQ-message will be embedded into a RANAP: INIT_UE_MESS which in turn will be embedded into an SCCP: CR-message. The SGSN will confirm SCCP-connection establishment by sending an SCCP: CC-message to the RNC. Optionally, the SCCP: CR- and CC-messages for connection establishment may be exchanged prior to transmitting the INIT_UE_MESS [ATT_REQ]-message in an SCCP: DT1-message.

> **Note:** The following section is optional. It will be conducted:
> - if the UE has not identified itself in ATT_REQ through its IMSI.
>
> and
> - if the SGSN cannot identify the UE based on its P-TMSI (neither in the VLR-database nor by possibly invoking the UE's authentication information from the previous SGSN.

⇒ The SGSN will request the UE to transmit its IMSI by sending an IDENT_REQ-message to the UE.
⇒ The UE will reply by sending its IMSI to the SGSN in an IDENT_RSP-message.

⇒ If necessary, the SGSN will invoke the UE's authentication quintet from the HLR or from the previous SGSN, before the AUTH_CIPH_REQ-message is sent to the UE.

(4) Attachment

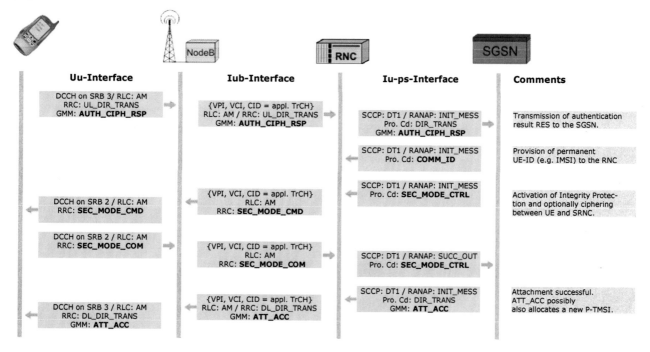

(4) Attachment

Description

⇒ The UE will calculate the authentication result RES and transmit it back to the SGSN in an AUTH_CIPH_RSP-message.
⇒ At this time, the SGSN will provide the permanent NAS (Non-Access-Stratum) UE-identity to the RNC (e.g. IMSI).
⇒ The SGSN will use the RANAP: SEC_MODE_CTRL-message to convey setup information related to ciphering and integrity protection towards the RNC. In turn, the RNC will an RRC: SEC_MODE_CMD-message towards the UE to start the integrity protection procedure and optionally encryption.
⇒ The UE shall confirm start of the integrity protection procedure and of ciphering by sending an RRC: SEC_MODE_COM-message to the RNC.
⇒ The RNC confirms the successful outcome of the RANAP-security mode control procedure by sending the RANAP: SEC_MODE_CTRL-message to the SGSN.
⇒ Finally, the SGSN will send the ATT_ACC-message to the UE. This message confirms the registration of the UE towards the packet-switched core network and may also contain a new P-TMSI.
⇒ If in NOM I, the ATT_ACC-message may also confirm circuit-switched IMSI-attachment and it may allocate a new TMSI.

(5) Attachment

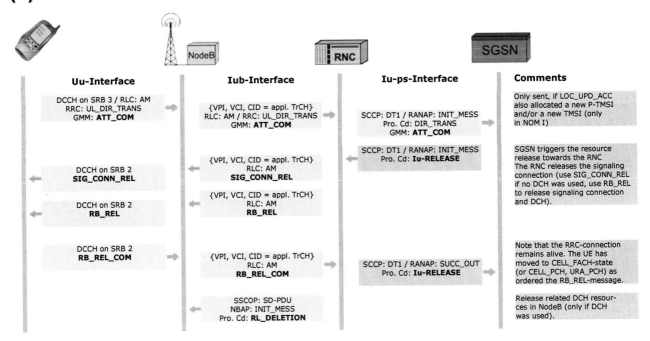

(5) Attachment

Description
⇒ Only if the ATT_ACC-message also allocated a new P-TMSI and/or TMSI to the UE, the UE shall confirm the new P-TMSI / TMSI by sending an ATT_COM-message to the SGSN.
⇒ The SGSN will trigger the resource release on all interfaces by sending an Iu-RELEASE-message to the RNC.
⇒ The SRNC intends to keep the RRC-connection to the UE alive. Accordingly, the SRNC will:
Possibility 1: Send an RRC: RB_REL-message to the UE, if a DCH was used for the Attach-procedure.
Possibility 2: Send an RRC: SIG_CONN_REL-message to the UE, if no DCH was used for the Attach-procedure.
In either case, the respective message indicates the RRC-connected state which the UE shall apply.
⇒ The UE will confirm the release of the signaling connection / of the radio bearer by sending the respective response message to the SRNC.
⇒ If a DCH was used, the RNC will send an RL_DELETION-message to the NodeB to initiate the release of the dedicated TrCH and of the physical resources between NodeB and UE.
⇒ The RNC will also confirm the release of the resources on the Iu-ps-interface by sending a RANAP: Iu-RELEASE-message (successful outcome) back to the SGSN.

(6) Attachment

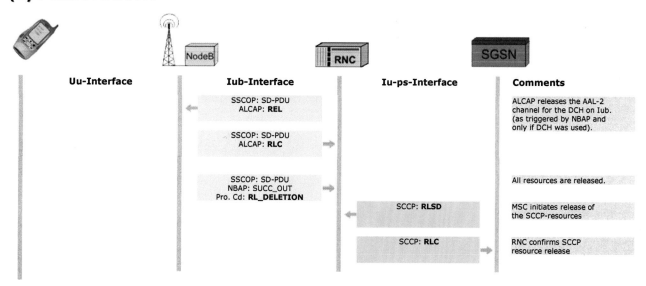

(6) Attachment

Description

⇒ If a DCH was used for the attachment, the RNC will initiate the release of the related AAL-2 channel on the Iub-interface (ALCAP: REL / RLC).
⇒ In this case, the NodeB will confirm the release of the dedicated TrCH and of the physical resources by sending an RL_DELETION-message (successful outcome) back the RNC.
⇒ The SCCP-connection is released by the SGSN by sending an SCCP: RLSD-message to the RNC. The RNC-confirms by sending an RLC-message back to the SGSN. The SCCP-connection release will most likely occur already after the reception of the Iu-RELEASE-message (successful outcome) by the SGSN.

(1) PDP-Context Activation (Mobile Originating)

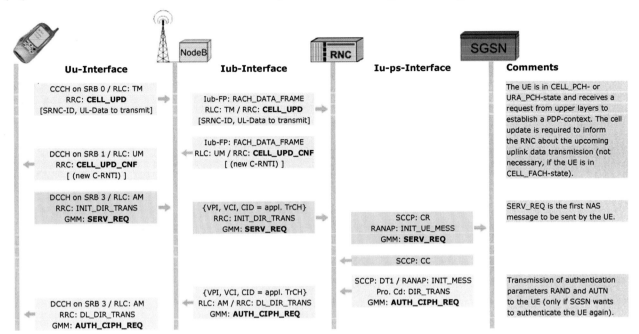

PDP context:
Needed for Packet Data Transfer

(1) PDP-Context Activation (Mobile Originating)

Initial Conditions
The UE is already RRC-connected (CELL_FACH-, CELL_PCH- or URA_PCH-state). The UE has previously attached to the packet-switched core network domain.

Applicability of this Procedure
This procedure is applicable in case of mobile originating PDP-context activation.

Description

> **Note:** The following procedure is optional. It will be conducted,
> • If the UE is in CELL_PCH- or URA-PCH-state and needs to move to CELL_FACH-state for uplink data transmission.

⇒ The UE will start the procedure by conducting a cell update scenario. As a consequence of the cell update scenario, the UE moves into CELL_FACH-state. In the presented case, no DCH's shall be configured by the SRNC.

⇒ The UE will transmit a GMM: SERV_REQ-message to the SRNC that will forward it to the SGSN in a RANAP: INIT_UE_MESS. This message is embedded in an SCCP: CR-message. The SGSN responds and confirms the SCCP-connection establishment by sending an SCCP: CC-message to the SRNC.

⇒ If configured by the network operator, the SGSN may authenticate the UE another time (was already done during attachment) by sending a GMM: AUTH_CIPH_REQ-message to the UE.

(2) PDP-Context Activation (Mobile Originating)

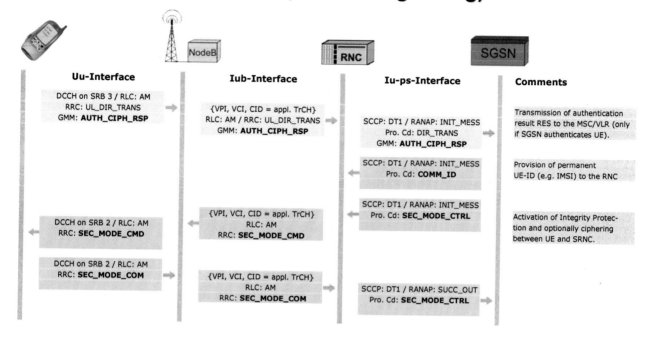

(2) PDP-Context Activation (Mobile Originating)

Description
⇒ The UE will calculate the authentication result RES and transmit it back to the SGSN in an AUTH_CIPH_RSP-message.
⇒ At this time, the SGSN will provide the permanent NAS (Non-Access-Stratum) UE-identity to the RNC (e.g. IMSI).
⇒ The SGSN will use the RANAP: SEC_MODE_CTRL-message to convey setup information related to ciphering and integrity protection towards the RNC. In turn, the RNC will an RRC: SEC_MODE_CMD-message towards the UE to start the integrity protection procedure and optionally encryption.
⇒ The UE shall confirm start of the integrity protection procedure and of ciphering by sending an RRC: SEC_MODE_COM-message to the RNC.
⇒ The RNC confirms the successful outcome of the RANAP-security mode control procedure by sending the RANAP: SEC_MODE_CTRL-message to the SGSN.

(3) PDP-Context Activation (Mobile Originating)

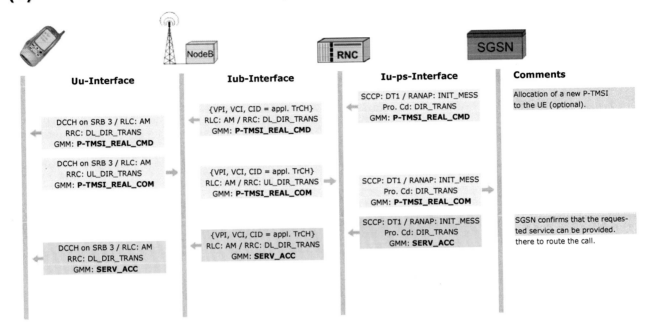

(3) PDP-Context Activation (Mobile Originating)

Description
- If configured by the network operator, the SGSN may reallocate the P-TMSI by sending a P-TMSI_REAL_CMD-message to the UE. The UE shall confirm by sending a P-TMSI_REAL_COM-message to the SGSN.
- If the SGSN is willing to accept the UE's service request, it shall send a GMM: SERV_REQ-message to the UE.

(4) PDP-Context Activation (Mobile Originating)

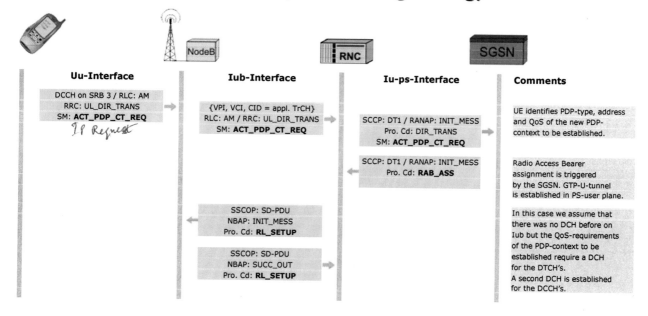

(4) PDP-Context Activation (Mobile Originating)

Description
⇒ Finally, the UE will transmit its SM: ACT_PDP_CT_REQ-message to the SGSN. This message identifies the requested QoS, the PDP-type and whether the UE requires the allocation of a dynamic PDP-address (most likely IP-address).
⇒ Now the SGSN will initiate the radio bearer assignment procedures by sending a RANAP: RAB_ASS-message (initiating message) to the SRNC. If possible, the allocated radio bearers should match the subscribed and / or requested QoS-profile.
⇒ Accordingly, the SRNC will initiate the NBAP: RL_SETUP-procedure on Iub- and Uu-interface. The NodeB confirms activation of the requested physical and transport channels by sending an NBAP: RL_SETUP-message (successful outcome) back to the SRNC.

(5) PDP-Context Activation (Mobile Originating)

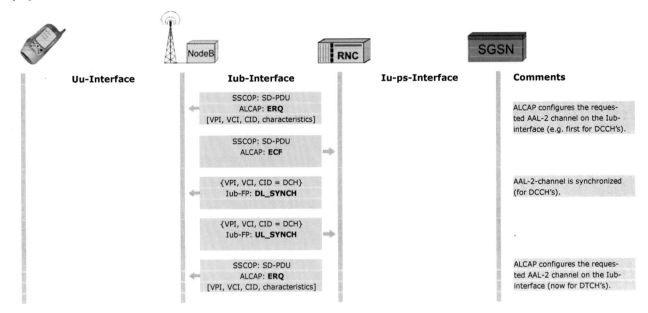

(5) PDP-Context Activation (Mobile Originating)

Description

Note that we illustrate a case when a DCH-TrCH is required for the PDP-context, because of the QoS as requested by the UE. In such a case, two DCH's are equipped for the DCCH's and for the DTCH's to provide for soft handover capabilities.

⇒ Finally, the UE will transmit its SM: ACT_PDP_CT_REQ-message to the SGSN. This message identifies the requested QoS, the PDP-type and whether the UE requires the allocation of a dynamic PDP-address (most likely IP-address).

(6) PDP-Context Activation (Mobile Originating)

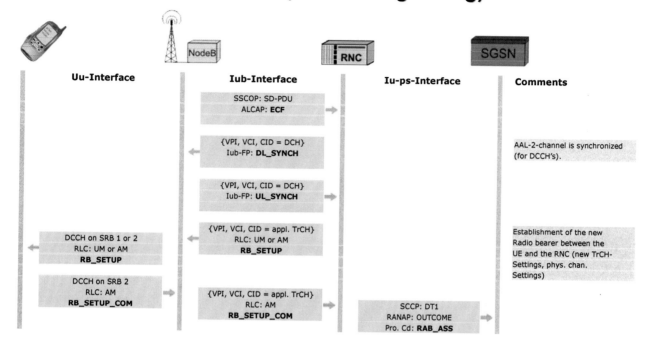

(6) PDP-Context Activation (Mobile Originating)

Description
⇒ Through ALCAP, the SRNC will establish AAL-2-links between itself and the NodeB. On Iu-ps-interface, a GTP-U-tunnel is being established.
⇒ On Iub-interface, the Iub-FP also synchronizes the new AAL-2 channel.
⇒ The SRNC will convey the new radio bearer setup configuration to the UE by sending an RRC: RB_SETUP-message to the UE.
⇒ The UE confirms by replying an RRC: RB_SETUP_COM-message to the SRNC.

(7) PDP-Context Activation (Mobile Originating)

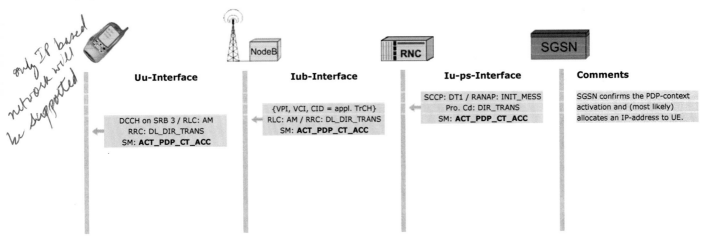

only IP based network will be supported

(7) PDP-Context Activation (Mobile Originating)

Description
⇒ The SGSN confirms the PDP-context activation by sending an SM: PDP_CT_ACT_ACC-message to the UE. This message most likely also contains the dynamic IP-address to be used by the UE.

Call Tracing in UTRAN

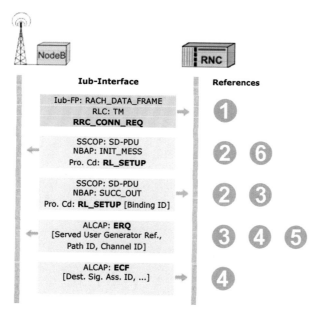

Call Tracing in UTRAN

Introduction
One of the most important tasks of various technicians is the recording file analysis in real life networks or in the laboratory to detect inconsistencies and errors. Obviously, this requires that the various messages in the different UMTS planes can be related to each other. The graphics page illustrates how this needs to be done for the messages on the Iub-interface. The different numbers underneath relate to the numbers on the graphics page.

Related Identifiers

1. **RRC-Information Element: Initial UE-Identity**
   The *Initial UE-Identity* (TMSI, P-TMSI, IMSI, ...) can be found in the RRC_CONN_REQ- and RRC_CONN_SETUP-message. Please note that the RRC_CONN_REQ-message will be sent on RACH while the RRC_CONN_SETUP-message is sent on FACH.

2. **NBAP-Information Element: CRNC-Communication Context ID (20 bit) and NodeB-Communication Context ID (20 bit)**
   a) CRNC-Communication Context ID
   For each new transaction the CRNC allocates the *CRNC-Communication Context ID*. This identifier will be unique for all NBAP-messages that the CRNC sends to the NodeB for one UE while the UE is having active links within this NodeB. In softer handover situations <u>no</u> new *CRNC-Communication Context ID* <u>nor</u> *NodeB-Communication Context ID* is used.
   b) NodeB-Communication Context ID
   When a new radio link is setup, the NodeB will send its response message (NBAP: RL_SETUP (successful outcome)) to the CRNC. This message will include the *NodeB-Communication Context ID* and the *CRNC-Communication Context ID*.

   Note: Usually, the *CRNC-Communication Context ID* and the *NodeB-Communication Context ID* are identical. They can be used to relate all NBAP-messages of a single transaction of one UE to each other.

3. **NBAP- and ALCAP-Information Elements: Binding ID and Served User Generator Reference**
   During the setup of a new radio link in the user or control plane (RL_SETUP-procedure), the NBAP-protocol in the NodeB allocates the *Binding ID's* that will be used by the ALCAP-protocol in the CRNC as *Served User Generator References*. By using these two identifiers, a new AAL-2-link can be related to the respective RL_SETUP-procedure.

4. **ALCAP Information Elements: Originating Signaling Association ID and Destination Signaling Association ID**
   ALCAP-messages can be related to each other through the *Originating Signaling Association ID* and *Destination Signaling Association ID*. The initial ERQ-message contains the Originating Signaling Association ID (from the sender's perspective). The receiver of this message will send an ALCAP: ECF-message back which contains its own Originating Signaling Association ID and relates back to the other Originating Signaling Association ID as Destination Signaling Association ID.

Call Tracing in UTRAN

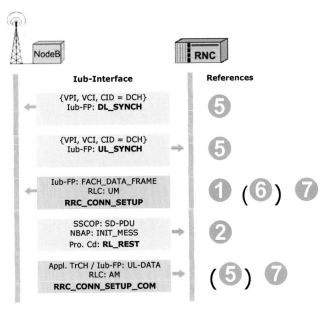

Call Tracing in UTRAN

5. **AAL-2 Link Channel ID and Path ID**
   If dedicated TrCH's shall be used for the DCCH's, then the AAL-2 path and channel ID's can be identified in the ALCAP: ERQ-message. They will remain unique for the lifetime of the transaction or until a possible channel reconfiguration.
   The 5- bullets are shown in brackets, because dedicated TrCH's for DCCH's (⇔ DCH's) may not necessarily have been configured. If no DCH's are used, then the identification of RRC-messages on DCCH's for one UE on the Iub-interface can be based on the RRC-transaction identifier, the RNTI or the RLC-sequence numbers (increasing per logical channel).
6. **The Uplink Scrambling Code**
   When an RRC_CONN_REQ-message is responded to by the SRNC, the respective RRC_CONN_SETUP-message is sent on FACH to the UE. To relate it back to the original RRC_CONN_REQ-message, the IE *Initial UE-Identity* has to be used. However, to relate the RRC_CONN_SETUP-message to the respective NBAP: RL_SETUP-message (initiating message) and the entire radio link setup procedure, the unique uplink scrambling code information element has to be used which can be found in both messages.
7. **RRC-Transaction Identifier / RNTI / RLC-Sequence Numbers**
   Each RRC-transaction (e.g. RRC_CONN_SETUP ⇔ RRC_CONN_SETUP_COM , ...) is guarded by the RRC-transaction identifier. It is also possible to use the C-RNTI in the MAC-header of common TrCH's to relate the RRC-messages from and to one UE to each other on the Iub-interface. Another means is to use the RLC-sequence numbers which are however separately counted on each DCCH.

UMTS – Signaling & Protocol Analysis

Practical Exercise:

- Determine the messages that belong together and put their numbers into the scenario underneath

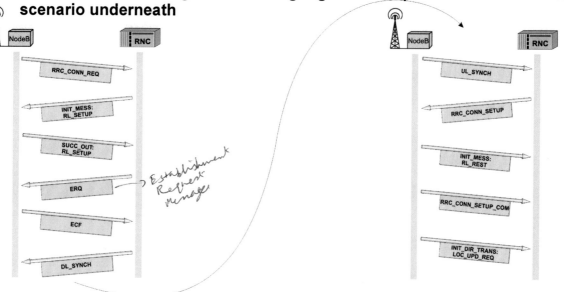

UMTS – Signaling & Protocol Analysis

Practical Exercise

- You need to use enclosure 1 to resolve this task. Enclosure 1 represents a typical recording file with different simultaneous scenarios.
- Apply the rules from the previous two slides to find the related messages.

Enclosures for the Practical Exercises

- Part 5: Important UMTS Scenarios & Call Tracing
- Exercise: Determine the messages that belong together and put their numbers into the scenario underneath.
  The following numbered messages represent a recording file. You shall insert the related numbers in the correct sequence into the solution scenario.

```
+----------+----------------------------------------+------------------------------------+
|BITMASK   |ID Name                                 |Comment or Value                    |
+----------+----------------------------------------+------------------------------------+
|TS 25.322 V3.10.0 (2002-03) reassembled (RLC reasm.)  TM DATA RACH (= Transparent Mode Data, Non Segmented, RACH)  |
|Transparent Mode Data, Non Segmented, RACH
|          |FP:  VPI/VCI/CID                        |"0/10/11"                           |
|          |FP:  Direction                          |Uplink                              |
|          |FP:  Transport Channel Type             |RACH (Random Access Channel)        |
|          |MAC: Target Channel Type                |CCCH (Common Control Channel)       |
|          |MAC: RLC Mode                           |Transparent Mode                    |
|**b166**  |RLC: Whole Data                         |'0010000100101110000000000000000101'B|
|          |                                        |1010100010011000100011101100000000'B|
|          |                                        |0000100000110000000000000000000000'B|
|          |                                        |0000000000000000000000000000000000'B|
|          |                                        |0000000000000000000000000000000000'B|
|          |                                        |                                    |
|TS 29.331 CCCH-UL (2002-03) (RRC_CCCH_UL)   rrcConnectionRequest (= rrcConnectionRequest)  |
|uL-CCCH-Message
|1 message
|1.1 rrcConnectionRequest
|1.1.1 initialUE-Identity
|1.1.1.1 tmsi-and-LAI
|***B4***  |1.1.1.1.1 tmsi                          |'00101110000000000000001011101010'B |
|1.1.1.1.2 lai
|1.1.1.1.2.1 plmn-Identity
|1.1.1.1.2.1.1 mcc
```

1

```
|0010----  |1.1.1.1.2.1.1.1 digit                   |2                                   |
|----0111  |1.1.1.1.2.1.1.2 digit                   |7                                   |
|0010----  |1.1.1.1.2.1.1.3 digit                   |2                                   |
|1.1.1.1.2.1.2 mnc
|***b4***  |1.1.1.1.2.1.2.1 digit                   |7                                   |
|-0110---  |1.1.1.1.2.1.2.2 digit                   |6                                   |
|**b16***  |1.1.1.1.2.2 lac                         |'0000000000010000'B                 |
|***b5***  |1.1.2 establishmentCause                |registration                        |
|--0-----  |1.1.3 protocolErrorIndicator            |noError                             |

+----------+----------------------------------------+------------------------------------+
|BITMASK   |ID Name                                 |Comment or Value                    |
+----------+----------------------------------------+------------------------------------+
|TS 25.322 V3.10.0 (2002-03) reassembled (RLC reasm.)  TM DATA RACH (= Transparent Mode Data, Non Segmented, RACH)  |
|Transparent Mode Data, Non Segmented, RACH
|          |FP:  VPI/VCI/CID                        |"0/10/11"                           |
|          |FP:  Direction                          |Uplink                              |
|          |FP:  Transport Channel Type             |RACH (Random Access Channel)        |
|          |MAC: Target Channel Type                |CCCH (Common Control Channel)       |
|          |MAC: RLC Mode                           |Transparent Mode                    |
|**b166**  |RLC: Whole Data                         |'0010001001011100000000000000000101'B|
|          |                                        |1010110010011000100011101100000000'B|
|          |                                        |0000100000000000000000000000000000'B|
|          |                                        |0000000000000000000000000000000000'B|
|          |                                        |0000000000000000000000000000000000'B|
|TS 29.331 CCCH-UL (2002-03) (RRC_CCCH_UL)   rrcConnectionRequest (= rrcConnectionRequest)  |
|uL-CCCH-Message
|1 message
|1.1 rrcConnectionRequest
|1.1.1 initialUE-Identity
|1.1.1.1 tmsi-and-LAI
|***B4***  |1.1.1.1.1 tmsi                          |'00111010000000000000001011101011'B |
|1.1.1.1.2 lai
|1.1.1.1.2.1 plmn-Identity
```

2

UMTS – Signaling & Protocol Analysis

```
|1.1.1.1.2.1.1 mcc
|0010----  |1.1.1.1.2.1.1.1 digit                       |2
|----0111  |1.1.1.1.2.1.1.2 digit                       |7
|0010----  |1.1.1.1.2.1.1.3 digit                       |2
|1.1.1.1.2.1.2 mnc
|***b4***  |1.1.1.1.2.1.2.1 digit                       |7
|-0110---  |1.1.1.1.2.1.2.2 digit                       |6
|**b16***  |1.1.1.1.2.2 lac                             |'0000000000010000'B
|***b5***  |1.1.2 establishmentCause                    |originatingConversationalCall
|--0-----  |1.1.3 protocolErrorIndicator                |noError


+----------+-----------------------------------------+---------------------------------------+
|BITMASK   |ID Name                                  |Comment or Value                       |
+----------+-----------------------------------------+---------------------------------------+
|UNI SSCOP (SSCOP)   SD (= Seq. Conn.mode Data)
|Seq. Conn.mode Data
|00000000  |Padding                                     |0
|01------  |PAD length                                  |1
|--00----  |Reserved                                    |0
|----1000  |PDU Type                                    |8
|***B3***  |this Sequencenumber - N(S)                  |116
|TS 25.433 V3.9.0 (2002-03) (NBAP)  initiatingMessage (= initiatingMessage)
|nbapPDU
|1 initiatingMessage
|1.1 procedureID
|00011011  |1.1.1 procedureCode                         |id-radioLinkSetup
|-01-----  |1.1.2 ddMode                                |fdd
|---00---  |1.2 criticality                             |reject
|-----0--  |1.3 messageDiscriminator                    |common
|1.4 transactionID
|***B2***  |1.4.1 longTransActionId                     |742
|1.5 value
|1.5.1 protocolIEs
|1.5.1.1 sequence
```

```
|***B2***  |1.5.1.1.1 id                                |id-CRNC-CommunicationContextID
|00------  |1.5.1.1.2 criticality                       |reject
|***B3***  |1.5.1.1.3 value                             |131814
|1.5.1.2 sequence
|***B2***  |1.5.1.2.1 id                                |id-UL-DPCH-Information-RL-SetupRqstF
|00------  |1.5.1.2.2 criticality                       |reject
|1.5.1.2.3 value
|1.5.1.2.3.1 ul-ScramblingCode
|***B3***  |1.5.1.2.3.1.1 uL-ScramblingCodeNumber       |1084105
|1-------  |1.5.1.2.3.1.2 uL-ScramblingCodeLength       |long
|--110---  |1.5.1.2.3.2 minUL-ChannelisationCodeLength  |v256
|***b4***  |1.5.1.2.3.3 ul-PunctureLimit                |15
|1.5.1.2.3.4 tFCS
|1.5.1.2.3.4.1 tFCSvalues
|1.5.1.2.3.4.1.1 no-Split-in-TFCI
|1.5.1.2.3.4.1.1.1 sequence
|1.5.1.2.3.4.1.1.1.1 cTFC
|------00  |1.5.1.2.3.4.1.1.1.1.1 ctfc2bit              |0
|1.5.1.2.3.4.1.1.1.2 tFC-Beta
|1.5.1.2.3.4.1.1.1.2.1 signalledGainFactors
|1.5.1.2.3.4.1.1.1.2.1.1 gainFactor
|1.5.1.2.3.4.1.1.1.2.1.1.1 fdd
|-1111---  |1.5.1.2.3.4.1.1.1.2.1.1.1.1 betaC           |15
|***b4***  |1.5.1.2.3.4.1.1.1.2.1.1.1.2 betaD           |10
|-00-----  |1.5.1.2.3.4.1.1.1.2.1.2 refTFCNumber        |0
|1.5.1.2.3.4.1.1.2 sequence
|1.5.1.2.3.4.1.1.2.1 cTFC
|-01-----  |1.5.1.2.3.4.1.1.2.1.1 ctfc2bit              |1
|1.5.1.2.3.4.1.1.2.2 tFC-Beta
|1.5.1.2.3.4.1.1.2.2.1 signalledGainFactors
|1.5.1.2.3.4.1.1.2.2.1.1 gainFactor
|1.5.1.2.3.4.1.1.2.2.1.1.1 fdd
|----1111  |1.5.1.2.3.4.1.1.2.2.1.1.1.1 betaC           |15
|1111----  |1.5.1.2.3.4.1.1.2.2.1.1.1.2 betaD           |15
|----00--  |1.5.1.2.3.4.1.1.2.2.1.2 refTFCNumber        |0
```

UMTS – Signaling & Protocol Analysis

```
|***b3***  |1.5.1.2.3.5 ul-DPCCH-SlotFormat             |0
|10001110  |1.5.1.2.3.6 ul-SIR-Target                   |60
|-00-----  |1.5.1.2.3.7 diversityMode                   |none
|1.5.1.3 sequence
|***B2***  |1.5.1.3.1 id                                |id-DL-DPCH-Information-RL-SetupRqstF
|00------  |1.5.1.3.2 criticality                       |reject
|1.5.1.3.3 value
|1.5.1.3.3.1 tFCS
|1.5.1.3.3.1.1 tFCSvalues
|1.5.1.3.3.1.1.1 no-Split-in-TFCI
|1.5.1.3.3.1.1.1.1 sequence
|1.5.1.3.3.1.1.1.1.1 cTFC
|------00  |1.5.1.3.3.1.1.1.1.1.1 ctfc2bit              |0
|1.5.1.3.3.1.1.1.2 sequence
|1.5.1.3.3.1.1.1.2.1 cTFC
|------01  |1.5.1.3.3.1.1.1.2.1.1 ctfc2bit              |1
|-00100--  |1.5.1.3.3.2 dl-DPCH-SlotFormat              |4
|1.5.1.3.3.3 tFCI-SignallingMode
|--0-----  |1.5.1.3.3.3.1 tFCI-SignallingOption         |normal
|---0----  |1.5.1.3.3.4 multiplexingPosition            |fixed
|1.5.1.3.3.5 powerOffsetInformation
|***b5***  |1.5.1.3.3.5.1 pO1-ForTFCI-Bits              |0
|---00000  |1.5.1.3.3.5.2 pO2-ForTPC-Bits               |0
|00000---  |1.5.1.3.3.5.3 pO3-ForPilotBits              |0
|------01  |1.5.1.3.3.6 fdd-TPC-DownlinkStepSize        |step-size1
|1-------  |1.5.1.3.3.7 limitedPowerIncrease            |not-used
|-0------  |1.5.1.3.3.8 innerLoopDLPCStatus             |active
|1.5.1.4 sequence
|***B2***  |1.5.1.4.1 id                                |id-DCH-FDD-Information
|00------  |1.5.1.4.2 criticality                       |reject
|1.5.1.4.3 value
|1.5.1.4.3.1 dCH-FDD-InformationItem
|--0-----  |1.5.1.4.3.1.1 payloadCRC-PresenceIndicator  |cRC-Included
|----1---  |1.5.1.4.3.1.2 ul-FP-Mode                    |silent
|***B2***  |1.5.1.4.3.1.3 toAWS                         |52
```

UMTS – Signaling & Protocol Analysis

```
|***B2***  |1.5.1.4.3.1.4 toAWE                             |1
|1.5.1.4.3.1.5 dCH-SpecificInformationList
|1.5.1.4.3.1.5.1 dCH-Specific-FDD-Item
|00011111  |1.5.1.4.3.1.5.1.1 dCH-ID                        |31
|1.5.1.4.3.1.5.1.2 ul-TransportFormatSet
|1.5.1.4.3.1.5.1.2.1 dynamicParts
|1.5.1.4.3.1.5.1.2.1.1 sequence
|***B2***  |1.5.1.4.3.1.5.1.2.1.1.1 nrOfTransportBlocks     |0
|1.5.1.4.3.1.5.1.2.1.1.2 mode
|          |1.5.1.4.3.1.5.1.2.1.1.2.1 notApplicable         |0
|1.5.1.4.3.1.5.1.2.1.2 sequence
|***B2***  |1.5.1.4.3.1.5.1.2.1.2.1 nrOfTransportBlocks     |1
|***B2***  |1.5.1.4.3.1.5.1.2.1.2.2 transportBlockSize      |148
|1.5.1.4.3.1.5.1.2.1.2.3 mode
|          |1.5.1.4.3.1.5.1.2.1.2.3.1 notApplicable         |0
|1.5.1.4.3.1.5.1.2.2 semi-staticPart
|***b3***  |1.5.1.4.3.1.5.1.2.2.1 transmissionTimeInter...  |msec-40
|--01----  |1.5.1.4.3.1.5.1.2.2.2 channelCoding             |convolutional-coding
|-----1--  |1.5.1.4.3.1.5.1.2.2.3 codingRate                |third
|10111000  |1.5.1.4.3.1.5.1.2.2.4 rateMatcingAttribute      |185
|-011----  |1.5.1.4.3.1.5.1.2.2.5 cRC-Size                  |v16
|1.5.1.4.3.1.5.1.2.2.6 mode
|          |1.5.1.4.3.1.5.1.2.2.6.1 notApplicable           |0
|1.5.1.4.3.1.5.1.3 dl-TransportFormatSet
|1.5.1.4.3.1.5.1.3.1 dynamicParts
|1.5.1.4.3.1.5.1.3.1.1 sequence
|***B2***  |1.5.1.4.3.1.5.1.3.1.1.1 nrOfTransportBlocks     |0
|1.5.1.4.3.1.5.1.3.1.1.2 mode
|          |1.5.1.4.3.1.5.1.3.1.1.2.1 notApplicable         |0
|1.5.1.4.3.1.5.1.3.1.2 sequence
|***B2***  |1.5.1.4.3.1.5.1.3.1.2.1 nrOfTransportBlocks     |1
|***B2***  |1.5.1.4.3.1.5.1.3.1.2.2 transportBlockSize      |148
|1.5.1.4.3.1.5.1.3.1.2.3 mode
|          |1.5.1.4.3.1.5.1.3.1.2.3.1 notApplicable         |0
|1.5.1.4.3.1.5.1.3.2 semi-staticPart
```

UMTS – Signaling & Protocol Analysis

```
|***b3***  |1.5.1.4.3.1.5.1.3.2.1 transmissionTimeInter..|msec-40
|--01----  |1.5.1.4.3.1.5.1.3.2.2 channelCoding          |convolutional-coding
|-----1--  |1.5.1.4.3.1.5.1.3.2.3 codingRate             |third
|10111000  |1.5.1.4.3.1.5.1.3.2.4 rateMatcingAttribute   |185
|-011----  |1.5.1.4.3.1.5.1.3.2.5 cRC-Size               |v16
|1.5.1.4.3.1.5.1.3.2.6 mode
|          |1.5.1.4.3.1.5.1.3.2.6.1 notApplicable        |0
|1.5.1.4.3.1.5.1.4 allocationRetentionPriority
|1111----  |1.5.1.4.3.1.5.1.4.1 priorityLevel            |15
|----0---  |1.5.1.4.3.1.5.1.4.2 pre-emptionCapability    |shall-not-trigger-pre-emption
|-----0--  |1.5.1.4.3.1.5.1.4.3 pre-emptionVulnerability |not-pre-emptable
|***b4***  |1.5.1.4.3.1.5.1.5 frameHandlingPriority      |0
|--0-----  |1.5.1.4.3.1.5.1.6 qE-Selector                |selected
|1.5.1.5 sequence
|***B2***  |1.5.1.5.1 id                                 |id-RL-InformationList-RL-SetupRqstFD
|10------  |1.5.1.5.2 criticality                        |notify
|1.5.1.5.3 value
|1.5.1.5.3.1 sequence
|***B2***  |1.5.1.5.3.1.1 id                             |id-RL-InformationItem-RL-SetupRqstFD
|10------  |1.5.1.5.3.1.2 criticality                    |notify
|1.5.1.5.3.1.3 value
|***b5***  |1.5.1.5.3.1.3.1 rL-ID                        |0
|***B2***  |1.5.1.5.3.1.3.2 c-ID                         |451
|-0------  |1.5.1.5.3.1.3.3 firstRLS-indicator           |first-RLS
|00000100  |1.5.1.5.3.1.3.4 frameOffset                  |4
|***B2***  |1.5.1.5.3.1.3.5 chipOffset                   |30720
|00000000  |1.5.1.5.3.1.3.6 propagationDelay             |0
|1.5.1.5.3.1.3.7 dl-CodeInformation
|1.5.1.5.3.1.3.7.1 fDD-DL-CodeInformationItem
|***b4***  |1.5.1.5.3.1.3.7.1.1 dl-ScramblingCode        |0
|***B2***  |1.5.1.5.3.1.3.7.1.2 fdd-DL-ChannelisationCo..|4
|***B2***  |1.5.1.5.3.1.3.8 initialDL-transmissionPower  |-100
|***B2***  |1.5.1.5.3.1.3.9 maximumDL-power              |-46
|***B2***  |1.5.1.5.3.1.3.10 minimumDL-power             |-296
```

UMTS – Signaling & Protocol Analysis

```
+---------+-----------------------------------------+----------------------------------+
|BITMASK  |ID Name                                  |Comment or Value                  |
+---------+-----------------------------------------+----------------------------------+
|UNI SSCOP (SSCOP)   SD (= Seq. Conn.mode Data)
|Seq. Conn.mode Data
|00000000 |Padding                                  |0
|01------ |PAD length                               |1
|--00---- |Reserved                                 |0
|----1000 |PDU Type                                 |8
|***B3*** |this Sequencenumber - N(S)               |112
|TS 25.433 V3.9.0 (2002-03) (NBAP)  initiatingMessage (= initiatingMessage)
|nbapPDU
|1 initiatingMessage
|1.1 procedureID
|00011011 |1.1.1 procedureCode                      |id-radioLinkSetup
|-01----- |1.1.2 ddMode                             |fdd
|---00--- |1.2 criticality                          |reject
|------0- |1.3 messageDiscriminator                 |common
|1.4 transactionID
|***B2*** |1.4.1 longTransActionId                  |741
|1.5 value
|1.5.1 protocolIEs
|1.5.1.1 sequence
|***B2*** |1.5.1.1.1 id                             |id-CRNC-CommunicationContextID
|00------ |1.5.1.1.2 criticality                    |reject
|***B3*** |1.5.1.1.3 value                          |131813
|1.5.1.2 sequence
|***B2*** |1.5.1.2.1 id                             |id-UL-DPCH-Information-RL-SetupRqstF
|00------ |1.5.1.2.2 criticality                    |reject
|1.5.1.2.3 value
|1.5.1.2.3.1 uL-ScramblingCode
|***B3*** |1.5.1.2.3.1.1 uL-ScramblingCodeNumber    |1084104
|1------- |1.5.1.2.3.1.2 uL-ScramblingCodeLength    |long
|--110--- |1.5.1.2.3.2 minUL-ChannelisationCodeLength|v256
```

UMTS – Signaling & Protocol Analysis

```
|***b4*** |1.5.1.2.3.3 ul-PunctureLimit              |15
|1.5.1.2.3.4 tFCS
|1.5.1.2.3.4.1 tFCSvalues
|1.5.1.2.3.4.1.1 no-Split-in-TFCI
|1.5.1.2.3.4.1.1.1 sequence
|1.5.1.2.3.4.1.1.1.1 cTFC
|------00 |1.5.1.2.3.4.1.1.1.1.1 ctfc2bit            |0
|1.5.1.2.3.4.1.1.1.2 tFC-Beta
|1.5.1.2.3.4.1.1.1.2.1 signalledGainFactors
|1.5.1.2.3.4.1.1.1.2.1.1 gainFactor
|1.5.1.2.3.4.1.1.1.2.1.1.1 fdd
|-1111--- |1.5.1.2.3.4.1.1.1.2.1.1.1.1 betaC          |15
|***b4*** |1.5.1.2.3.4.1.1.1.2.1.1.1.2 betaD          |10
|-00----- |1.5.1.2.3.4.1.1.1.2.1.2 refTFCNumber       |0
|1.5.1.2.3.4.1.1.2 sequence
|1.5.1.2.3.4.1.1.2.1 cTFC
|-01----- |1.5.1.2.3.4.1.1.2.1.1 ctfc2bit             |1
|1.5.1.2.3.4.1.1.2.2 tFC-Beta
|1.5.1.2.3.4.1.1.2.2.1 signalledGainFactors
|1.5.1.2.3.4.1.1.2.2.1.1 gainFactor
|1.5.1.2.3.4.1.1.2.2.1.1.1 fdd
|----1111 |1.5.1.2.3.4.1.1.2.2.1.1.1.1 betaC          |15
|1111---- |1.5.1.2.3.4.1.1.2.2.1.1.1.2 betaD          |15
|----00-- |1.5.1.2.3.4.1.1.2.2.1.2 refTFCNumber       |0
|***b3*** |1.5.1.2.3.5 ul-DPCCH-SlotFormat            |0
|10001110 |1.5.1.2.3.6 ul-SIR-Target                  |60
|-00----- |1.5.1.2.3.7 diversityMode                  |none
|1.5.1.3 sequence
|***B2*** |1.5.1.3.1 id                               |id-DL-DPCH-Information-RL-SetupRqstF
|00------ |1.5.1.3.2 criticality                      |reject
|1.5.1.3.3 value
|1.5.1.3.3.1 tFCS
|1.5.1.3.3.1.1 tFCSvalues
|1.5.1.3.3.1.1.1 no-Split-in-TFCI
|1.5.1.3.3.1.1.1.1 sequence
```

UMTS – Signaling & Protocol Analysis

```
|1.5.1.3.3.1.1.1.1.1 cTFC
|------00 |1.5.1.3.3.1.1.1.1.1.1 ctfc2bit            |0
|1.5.1.3.3.1.1.1.2 sequence
|1.5.1.3.3.1.1.1.2.1 cTFC
|------01 |1.5.1.3.3.1.1.1.2.1.1 ctfc2bit            |1
|-00100-- |1.5.1.3.3.2 dl-DPCH-SlotFormat             |4
|1.5.1.3.3.3 tFCI-SignallingMode
|--0----- |1.5.1.3.3.3.1 tFCI-SignallingOption        |normal
|---0---- |1.5.1.3.3.4 multiplexingPosition           |fixed
|1.5.1.3.3.5 powerOffsetInformation
|***b5*** |1.5.1.3.3.5.1 pO1-ForTFCI-Bits             |0
|---00000 |1.5.1.3.3.5.2 pO2-ForTPC-Bits              |0
|00000--- |1.5.1.3.3.5.3 pO3-ForPilotBits             |0
|------01 |1.5.1.3.3.6 fdd-TPC-DownlinkStepSize       |step-size1
|1------- |1.5.1.3.3.7 limitedPowerIncrease           |not-used
|-0------ |1.5.1.3.3.8 innerLoopDLPCStatus            |active
|1.5.1.4 sequence
|***B2*** |1.5.1.4.1 id                               |id-DCH-FDD-Information
|00------ |1.5.1.4.2 criticality                      |reject
|1.5.1.4.3 value
|1.5.1.4.3.1 dCH-FDD-InformationItem
|--0----- |1.5.1.4.3.1.1 payloadCRC-PresenceIndicator |cRC-Included
|----1--- |1.5.1.4.3.1.2 ul-FP-Mode                   |silent
|***B2*** |1.5.1.4.3.1.3 toAWS                        |52
|***B2*** |1.5.1.4.3.1.4 toAWE                        |1
|1.5.1.4.3.1.5 dCH-SpecificInformationList
|1.5.1.4.3.1.5.1 dCH-Specific-FDD-Item
|00011111 |1.5.1.4.3.1.5.1.1 dCH-ID                   |31
|1.5.1.4.3.1.5.1.2 ul-TransportFormatSet
|1.5.1.4.3.1.5.1.2.1 dynamicParts
|1.5.1.4.3.1.5.1.2.1.1 sequence
|***B2*** |1.5.1.4.3.1.5.1.2.1.1.1 nrOfTransportBlocks|0
|1.5.1.4.3.1.5.1.2.1.1.2 mode
|         |1.5.1.4.3.1.5.1.2.1.1.2.1 notApplicable   |0
|1.5.1.4.3.1.5.1.2.1.2 sequence
```

UMTS – Signaling & Protocol Analysis

```
|***B2***  |1.5.1.4.3.1.5.1.2.1.2.1 nrOfTransportBlocks   |1
|***B2***  |1.5.1.4.3.1.5.1.2.1.2.2 transportBlockSize    |148
|1.5.1.4.3.1.5.1.2.1.2.3 mode
|          |1.5.1.4.3.1.5.1.2.1.2.3.1 notApplicable       |0
|1.5.1.4.3.1.5.1.2.2 semi-staticPart
|***b3***  |1.5.1.4.3.1.5.1.2.2.1 transmissionTimeInter.. |msec-40
|--01----  |1.5.1.4.3.1.5.1.2.2.2 channelCoding           |convolutional-coding
|-----1--  |1.5.1.4.3.1.5.1.2.2.3 codingRate              |third
|10111000  |1.5.1.4.3.1.5.1.2.2.4 rateMatcingAttribute    |185
|-011----  |1.5.1.4.3.1.5.1.2.2.5 cRC-Size                |v16
|1.5.1.4.3.1.5.1.2.2.6 mode
|          |1.5.1.4.3.1.5.1.2.2.6.1 notApplicable         |0
|1.5.1.4.3.1.5.1.3 dl-TransportFormatSet
|1.5.1.4.3.1.5.1.3.1 dynamicParts
|1.5.1.4.3.1.5.1.3.1.1 sequence
|***B2***  |1.5.1.4.3.1.5.1.3.1.1.1 nrOfTransportBlocks   |0
|1.5.1.4.3.1.5.1.3.1.1.2 mode
|          |1.5.1.4.3.1.5.1.3.1.1.2.1 notApplicable       |0
|1.5.1.4.3.1.5.1.3.1.2 sequence
|***B2***  |1.5.1.4.3.1.5.1.3.1.2.1 nrOfTransportBlocks   |1
|***B2***  |1.5.1.4.3.1.5.1.3.1.2.2 transportBlockSize    |148
|1.5.1.4.3.1.5.1.3.1.2.3 mode
|          |1.5.1.4.3.1.5.1.3.1.2.3.1 notApplicable       |0
|1.5.1.4.3.1.5.1.3.2 semi-staticPart
|***b3***  |1.5.1.4.3.1.5.1.3.2.1 transmissionTimeInter.. |msec-40
|--01----  |1.5.1.4.3.1.5.1.3.2.2 channelCoding           |convolutional-coding
|-----1--  |1.5.1.4.3.1.5.1.3.2.3 codingRate              |third
|10111000  |1.5.1.4.3.1.5.1.3.2.4 rateMatcingAttribute    |185
|-011----  |1.5.1.4.3.1.5.1.3.2.5 cRC-Size                |v16
|1.5.1.4.3.1.5.1.3.2.6 mode
|          |1.5.1.4.3.1.5.1.3.2.6.1 notApplicable         |0
|1.5.1.4.3.1.5.1.4 allocationRetentionPriority
|1111----  |1.5.1.4.3.1.5.1.4.1 priorityLevel             |15
|----0---  |1.5.1.4.3.1.5.1.4.2 pre-emptionCapability     |shall-not-trigger-pre-emption
|-----0--  |1.5.1.4.3.1.5.1.4.3 pre-emptionVulnerability  |not-pre-emptable
```

UMTS – Signaling & Protocol Analysis

```
|***b4***  |1.5.1.4.3.1.5.1.5 frameHandlingPriority       |0
|--0-----  |1.5.1.4.3.1.5.1.6 qE-Selector                 |selected
|1.5.1.5 sequence
|***B2***  |1.5.1.5.1 id                                  |id-RL-InformationList-RL-SetupRqstFD
|10------  |1.5.1.5.2 criticality                         |notify
|1.5.1.5.3 value
|1.5.1.5.3.1 sequence
|***B2***  |1.5.1.5.3.1.1 id                              |id-RL-InformationItem-RL-SetupRqstFD
|10------  |1.5.1.5.3.1.2 criticality                     |notify
|1.5.1.5.3.1.3 value
|***b5***  |1.5.1.5.3.1.3.1 rL-ID                         |0
|***B2***  |1.5.1.5.3.1.3.2 c-ID                          |451
|-0------  |1.5.1.5.3.1.3.3 firstRLS-indicator            |first-RLS
|00000100  |1.5.1.5.3.1.3.4 frameOffset                   |4
|***B2***  |1.5.1.5.3.1.3.5 chipOffset                    |31232
|00000000  |1.5.1.5.3.1.3.6 propagationDelay              |0
|1.5.1.5.3.1.3.7 dl-CodeInformation
|1.5.1.5.3.1.3.7.1 fDD-DL-CodeInformationItem
|***b4***  |1.5.1.5.3.1.3.7.1.1 dl-ScramblingCode         |0
|***B2***  |1.5.1.5.3.1.3.7.1.2 fdd-DL-ChannelisationCo.. |4
|***B2***  |1.5.1.5.3.1.3.8 initialDL-transmissionPower   |-100
|***B2***  |1.5.1.5.3.1.3.9 maximumDL-power               |-46
|***B2***  |1.5.1.5.3.1.3.10 minimumDL-power              |-296


+----------+-----------------------------------------------+-----------------------------------+
|BITMASK   |ID Name                                        |Comment or Value                   |
+----------+-----------------------------------------------+-----------------------------------+
|UNI SSCOP (SSCOP)   SD (= Seq. Conn.mode Data)
|Seq. Conn.mode Data
|00000000  |Padding                                        |0
|01------  |PAD length                                     |1
|--00----  |Reserved                                       |0
|----1000  |PDU Type                                       |8
|***B3***  |this Sequencenumber - N(S)                     |188
```

UMTS – Signaling & Protocol Analysis

```
|TS 25.433 V3.9.0 (2002-03) (NBAP)  succesfulOutcome (= succesfulOutcome)   |
|nbapPDU                                                                    |
|1 succesfulOutcome                                                         |
|1.1 procedureID                                                            |
|00011011  |1.1.1 procedureCode             |id-radioLinkSetup             |
|-01-----  |1.1.2 ddMode                    |fdd                           |
|---00---  |1.2 criticality                 |reject                        |
|-----0--  |1.3 messageDiscriminator        |common                        |
|1.4 transactionID                                                          |
|***B2***  |1.4.1 longTransActionId         |741                           |
|1.5 value                                                                  |
|1.5.1 protocolIEs                                                          |
|1.5.1.1 sequence                                                           |
|***B2***  |1.5.1.1.1 id                    |id-CRNC-CommunicationContextID|
|01------  |1.5.1.1.2 criticality           |ignore                        |
|***B3***  |1.5.1.1.3 value                 |131813                        |
|1.5.1.2 sequence                                                           |
|***B2***  |1.5.1.2.1 id                    |id-NodeB-CommunicationContextID|
|01------  |1.5.1.2.2 criticality           |ignore                        |
|***B3***  |1.5.1.2.3 value                 |131813                        |
|1.5.1.3 sequence                                                           |
|***B2***  |1.5.1.3.1 id                    |id-CommunicationControlPortID |
|01------  |1.5.1.3.2 criticality           |ignore                        |
|***B2***  |1.5.1.3.3 value                 |0                             |
|1.5.1.4 sequence                                                           |
|***B2***  |1.5.1.4.1 id                    |id-RL-InformationResponseList-RL-Set|
|01------  |1.5.1.4.2 criticality           |ignore                        |
|1.5.1.4.3 value                                                            |
|1.5.1.4.3.1 sequenceOf                                                     |
|***B2***  |1.5.1.4.3.1.1 id                |id-RL-InformationResponseItem-RL-Set|
|01------  |1.5.1.4.3.1.2 criticality       |ignore                        |
|1.5.1.4.3.1.3 value                                                        |
|---00000  |1.5.1.4.3.1.3.1 rL-ID           |0                             |
|00001---  |1.5.1.4.3.1.3.2 rL-Set-ID       |1                             |
|***B2***  |1.5.1.4.3.1.3.3 received-total-wide-band-po..|0                 |
```

UMTS – Signaling & Protocol Analysis

```
|1.5.1.4.3.1.3.4 diversityIndication                                        |
|1.5.1.4.3.1.3.4.1 nonCombiningOrFirstRL                                    |
|1.5.1.4.3.1.3.4.1.1 dCH-InformationResponse                                |
|1.5.1.4.3.1.3.4.1.1.1 dCH-InformationResponseItem                          |
|00011111  |1.5.1.4.3.1.3.4.1.1.1.1 dCH-ID       |31                        |
|***B4***  |1.5.1.4.3.1.3.4.1.1.1.2 bindingID    |24 00 00 00               |
|**B20***  |1.5.1.4.3.1.3.4.1.1.1.3 transportLayerAddress|(1), (3), (5), (6), (9), (10), (11),|
|1-------  |1.5.1.4.3.1.3.5 sSDT-SupportIndicator|sSDT-not-supported        |
```

```
+---------+------------------------------------+----------------------------+
|BITMASK  |ID Name                             |Comment or Value            |
+---------+------------------------------------+----------------------------+
|UNI SSCOP (SSCOP)   SD (= Seq. Conn.mode Data)                             |
|Seq. Conn.mode Data                                                        |
|00------ |PAD length                          |0                           |
|--00---- |Reserved                            |0                           |
|----1000 |PDU Type                            |8                           |
|***B3*** |this Sequencenumber - N(S)          |61                          |
|ITU-T Q.2630.1/2 AAL2 Signalling CS1/2 (AAL2L3)  ERQ (= Establish Request) |
|Establish Request                                                          |
|***B4*** |Dest. Sign Assoc. Id.               |0                           |
|00000101 |Message Identifier                  |5                           |
|------10 |Instr. ind. - general action        |Discard message             |
|-----0-- |Send not. ind. - general action     |Do not send notification    |
|----0--- |reserved - general action           |reserved                    |
|--10---- |Instr.ind. - pass-on not possible   |Discard message             |
|-0------ |Send not. ind. - pass-on not possible|Do not send notification   |
|0------- |reserved - pass-on not possible     |reserved                    |
|Orig Signalling Association Id                                             |
|00000110 |IE Name                             |Origin. Signall. Ass. Id.   |
|------10 |Instr. ind. - general action        |Discard message             |
|-----0-- |Send not. ind. - general action     |Do not send notification    |
|----0--- |reserved - general action           |reserved                    |
|--10---- |Instr.ind. - pass-on not possible   |Discard message             |
```

```
|00------ |Send not. ind. - pass-on not possible |Do not send notification      |
|00000100 |Length of contents of IE              |4                             |
|***B4*** |Originating signal. ass. Id.          |33555272                      |
|Connection Element Id                           |                              |
|00000010 |IE Name                               |conn. elem. id                |
|------10 |Instr. ind. - general action          |Discard message               |
|-----0-- |Send not. ind. - general action       |Do not send notification      |
|----0--- |reserved - general action             |reserved                      |
|--10---- |Instr.ind. - pass-on not possible     |Discard message               |
|00------ |Send not. ind. - pass-on not possible |Do not send notification      |
|00000101 |Length of contents of IE              |5                             |
|***B4*** |AAL2 type 2 path id.                  |67                            |
|11111101 |channel id. (0, 8-255)                |253                           |
|Dest NSAP  Serv. Endp. Address                  |                              |
|00000100 |IE Name                               |Dest NSAP Serv. EP Addr.      |
|------10 |Instr. ind. - general action          |Discard message               |
|-----0-- |Send not. ind. - general action       |Do not send notification      |
|----0--- |reserved - general action             |reserved                      |
|--10---- |Instr.ind. - pass-on not possible     |Discard message               |
|00------ |Send not. ind. - pass-on not possible |Do not send notification      |
|00010100 |Length of contents of IE              |20                            |
|**B20*** |NSAP_Address                          |45 56 78 9a bc de f0 12 0f 00 00... |
|Link Charakteristics                            |                              |
|00000101 |IE Name                               |Link Characteristics          |
|------10 |Instr. ind. - general action          |Discard message               |
|-----0-- |Send not. ind. - general action       |Do not send notification      |
|----0--- |reserved - general action             |reserved                      |
|--10---- |Instr.ind. - pass-on not possible     |Discard message               |
|00------ |Send not. ind. - pass-on not possible |Do not send notification      |
|00001100 |Length of contents of IE              |12                            |
|***B2*** |Max. CPS-SDU BR (FwD)                 |75                            |
|***B2*** |Max. CPS-SDU BR (BwD)                 |80                            |
|***B2*** |Avrg. CPS-SDU BR (FwD)                |15                            |
|***B2*** |Avrg. CPS-SDU BR (BwD)                |16                            |
|00011000 |Max. CPS-SDU Size (FwD)               |24                            |
```

```
|00011010 |Max. CPS-SDU Size (BwD)               |26                            |
|00011000 |Avrg. CPS-SDU Size (FwD)              |24                            |
|00011010 |Avrg. CPS-SDU Size (BwD)              |26                            |
|Served User Gen. Reference                      |                              |
|00000111 |IE Name                               |Served User Gen. Reference    |
|------10 |Instr. ind. - general action          |Discard message               |
|-----0-- |Send not. ind. - general action       |Do not send notification      |
|----0--- |reserved - general action             |reserved                      |
|--10---- |Instr.ind. - pass-on not possible     |Discard message               |
|00------ |Send not. ind. - pass-on not possible |Do not send notification      |
|00000100 |Length of contents of IE              |4                             |
|***B4*** |served user gen reference             |620756992                     |
|Served User Transport                           |                              |
|00001000 |IE Name                               |Served User Transport         |
|------01 |Instr. ind. - general action          |Discard parameter             |
|-----0-- |Send not. ind. - general action       |Do not send notification      |
|----0--- |reserved - general action             |reserved                      |
|--01---- |Instr.ind. - pass-on not possible     |Discard parameter             |
|00------ |Send not. ind. - pass-on not possible |Do not send notification      |
|00000011 |Length of contents of IE              |3                             |
|00000010 |Length of Served user Transport       |2                             |
|***B2*** |Served user Trans. in HEX             |04 27                         |

+---------+-------------------------------------+------------------------------+
|BITMASK  |ID Name                              |Comment or Value              |
+---------+-------------------------------------+------------------------------+
|UNI SSCOP (SSCOP)   SD (= Seq. Conn.mode Data)                                 |
|Seq. Conn.mode Data                                                            |
|00------ |PAD length                           |0                             |
|--00---- |Reserved                             |0                             |
|----1000 |PDU Type                             |8                             |
|***B3*** |this Sequencenumber - N(S)           |59                            |
|ITU-T Q.2630.1/2 AAL2 Signalling CS1/2 (AAL2L3)   ERQ (= Establish Request)    |
|Establish Request                                                              |
```

UMTS – Signaling & Protocol Analysis

```
|***B4***  |Dest. Sign Assoc. Id.                |0                        |
|00000101  |Message Identifier                   |5                        |
|------10  |Instr. ind. - general action         |Discard message          |
|-----0--  |Send not. ind. - general action      |Do not send notification |
|----0---  |reserved - general action            |reserved                 |
|--10----  |Instr.ind. - pass-on not possible    |Discard message          |
|-0------  |Send not. ind. - pass-on not possible|Do not send notification |
|0-------  |reserved - pass-on not possible      |reserved                 |
|Orig Signalling Association Id                                            |
|00000110  |IE Name                              |Origin. Signall. Ass. Id.|
|------10  |Instr. ind. - general action         |Discard message          |
|-----0--  |Send not. ind. - general action      |Do not send notification |
|----0---  |reserved - general action            |reserved                 |
|--10----  |Instr.ind. - pass-on not possible    |Discard message          |
|00------  |Send not. ind. - pass-on not possible|Do not send notification |
|00000100  |Length of contents of IE             |4                        |
|***B4***  |Originating signal. ass. Id.         |33555271                 |
|Connection Element Id                                                     |
|00000010  |IE Name                              |conn. elem. id           |
|------10  |Instr. ind. - general action         |Discard message          |
|-----0--  |Send not. ind. - general action      |Do not send notification |
|----0---  |reserved - general action            |reserved                 |
|--10----  |Instr.ind. - pass-on not possible    |Discard message          |
|00------  |Send not. ind. - pass-on not possible|Do not send notification |
|00000101  |Length of contents of IE             |5                        |
|***B4***  |AAL2 type 2 path id.                 |06                       |
|01000000  |channel id. (0, 8-255)               |64                       |
|Dest NSAP  Serv. Endp. Address                                            |
|00000100  |IE Name                              |Dest NSAP Serv. EP Addr. |
|------10  |Instr. ind. - general action         |Discard message          |
|-----0--  |Send not. ind. - general action      |Do not send notification |
|----0---  |reserved - general action            |reserved                 |
|--10----  |Instr.ind. - pass-on not possible    |Discard message          |
|00------  |Send not. ind. - pass-on not possible|Do not send notification |
|00010100  |Length of contents of IE             |20                       |
```

UMTS – Signaling & Protocol Analysis

```
|**B20***  |NSAP_Address                         |45 56 78 9a bc de f0 12 0f 00 00...|
|Link Charakteristics                                                      |
|00000101  |IE Name                              |Link Characteristics     |
|------10  |Instr. ind. - general action         |Discard message          |
|-----0--  |Send not. ind. - general action      |Do not send notification |
|----0---  |reserved - general action            |reserved                 |
|--10----  |Instr.ind. - pass-on not possible    |Discard message          |
|00------  |Send not. ind. - pass-on not possible|Do not send notification |
|00001100  |Length of contents of IE             |12                       |
|***B2***  |Max. CPS-SDU BR (FwD)                |75                       |
|***B2***  |Max. CPS-SDU BR (BwD)                |80                       |
|***B2***  |Avrg. CPS-SDU BR (FwD)               |15                       |
|***B2***  |Avrg. CPS-SDU BR (BwD)               |16                       |
|00011000  |Max. CPS-SDU Size (FwD)              |24                       |
|00011010  |Max. CPS-SDU Size (BwD)              |26                       |
|00011000  |Avrg. CPS-SDU Size (FwD)             |24                       |
|00011010  |Avrg. CPS-SDU Size (BwD)             |26                       |
|Served User Gen. Reference                                                |
|00000111  |IE Name                              |Served User Gen. Reference|
|------10  |Instr. ind. - general action         |Discard message          |
|-----0--  |Send not. ind. - general action      |Do not send notification |
|----0---  |reserved - general action            |reserved                 |
|--10----  |Instr.ind. - pass-on not possible    |Discard message          |
|00------  |Send not. ind. - pass-on not possible|Do not send notification |
|00000100  |Length of contents of IE             |4                        |
|***B4***  |served user gen reference            |603979776                |
|Served User Transport                                                     |
|00001000  |IE Name                              |Served User Transport    |
|------01  |Instr. ind. - general action         |Discard parameter        |
|-----0--  |Send not. ind. - general action      |Do not send notification |
|----0---  |reserved - general action            |reserved                 |
|--01----  |Instr.ind. - pass-on not possible    |Discard parameter        |
|00------  |Send not. ind. - pass-on not possible|Do not send notification |
|00000011  |Length of contents of IE             |3                        |
|00000010  |Length of Served user Transport      |2                        |
```

UMTS – Signaling & Protocol Analysis

```
|***B2***  |Served user Trans. in HEX              |04 27                              |

+---------+----------------------------------------+-----------------------------------+
|BITMASK  |ID Name                                 |Comment or Value                   |
+---------+----------------------------------------+-----------------------------------+
|UNI SSCOP (SSCOP)  SD (= Seq. Conn.mode Data)
|Seq. Conn.mode Data
|***B3*** |Padding                                 |0
|11------ |PAD length                              |3
|--00---- |Reserved                                |0
|----1000 |PDU Type                                |8
|***B3*** |this Sequencenumber - N(S)              |61
|ITU-T Q.2630.1/2 AAL2 Signalling CS1/2 (AAL2L3)  ECF (= Establish Confirm)
|Establish Confirm
|***B4*** |Dest. Sign Assoc. Id.                   |33555272
|00000100 |Message Identifier                      |4
|------10 |Instr. ind. - general action            |Discard message
|-----0-- |Send not. ind. - general action         |Do not send notification
|----0--- |reserved - general action               |reserved
|--10---- |Instr.ind. - pass-on not possible       |Discard message
|-0------ |Send not. ind. - pass-on not possible   |Do not send notification
|0------- |reserved - pass-on not possible         |reserved
|Orig Signalling Association Id
|00000110 |IE Name                                 |Origin. Signall. Ass. Id.
|------10 |Instr. ind. - general action            |Discard message
|-----0-- |Send not. ind. - general action         |Do not send notification
|----0--- |reserved - general action               |reserved
|--10---- |Instr.ind. - pass-on not possible       |Discard message
|00------ |Send not. ind. - pass-on not possible   |Do not send notification
|00000100 |Length of contents of IE                |4
|***B4*** |Originating signal. ass. Id.            |13
+---------+----------------------------------------+-----------------------------------+
```

 8

UMTS – Signaling & Protocol Analysis

```
|BITMASK  |ID Name                                 |Comment or Value                   |
+---------+----------------------------------------+-----------------------------------+
|UNI SSCOP (SSCOP)  SD (= Seq. Conn.mode Data)
|Seq. Conn.mode Data
|***B3*** |Padding                                 |0
|11------ |PAD length                              |3
|--00---- |Reserved                                |0
|----1000 |PDU Type                                |8
|***B3*** |this Sequencenumber - N(S)              |59
|ITU-T Q.2630.1/2 AAL2 Signalling CS1/2 (AAL2L3)  ECF (= Establish Confirm)
|Establish Confirm
|***B4*** |Dest. Sign Assoc. Id.                   |33555271
|00000100 |Message Identifier                      |4
|------10 |Instr. ind. - general action            |Discard message
|-----0-- |Send not. ind. - general action         |Do not send notification
|----0--- |reserved - general action               |reserved
|--10---- |Instr.ind. - pass-on not possible       |Discard message
|-0------ |Send not. ind. - pass-on not possible   |Do not send notification
|0------- |reserved - pass-on not possible         |reserved
|Orig Signalling Association Id
|00000110 |IE Name                                 |Origin. Signall. Ass. Id.
|------10 |Instr. ind. - general action            |Discard message
|-----0-- |Send not. ind. - general action         |Do not send notification
|----0--- |reserved - general action               |reserved
|--10---- |Instr.ind. - pass-on not possible       |Discard message
|00------ |Send not. ind. - pass-on not possible   |Do not send notification
|00000100 |Length of contents of IE                |4
|***B4*** |Originating signal. ass. Id.            |17

+---------+----------------------------------------+-----------------------------------+
|BITMASK  |ID Name                                 |Comment or Value                   |
+---------+----------------------------------------+-----------------------------------+
|TS 25.322 V3.10.0 (RLC) / 25.321 V3.11.0 (MAC) / 25.435 V3.10.0, 25.427 V3.9.0 (FP) - (2002-03)  (RLC/MAC)  FP CTRL DCH
|(= FP Control Frame DCH)  |
```

 9

10

UMTS – Signaling & Protocol Analysis

```
         |FP Control Frame DCH
         |        |FP:  VPI/VCI/CID                      |"14/06/64"
         |        |FP:  Radio Mode                       |FDD (Frequency Division Duplex)
         |        |FP:  Direction                        |Downlink
         |        |FP:  Transport Channel Type           |DCH (Dedicated Channel)
|0101010- |FP:  Frame CRC                                |2a
|-------1 |FP:  Frame Type                               |Control
|00000011 |FP:  Control Frame Type (DCH)                 |DL Synchronisation
|11110000 |FP:  Connection Frame Number                  |240


+---------+--------------------------------------+------------------------------------+
|BITMASK  |ID Name                               |Comment or Value                    |           11
+---------+--------------------------------------+------------------------------------+
|TS 25.322 V3.10.0 (RLC) / 25.321 V3.11.0 (MAC) / 25.435 V3.10.0, 25.427 V3.9.0 (FP) - (2002-03) (RLC/MAC)   FP CTRL DCH
(= FP Control Frame DCH)  |
|FP Control Frame DCH
         |        |FP:  VPI/VCI/CID                      |"14/06/64"
         |        |FP:  Radio Mode                       |FDD (Frequency Division Duplex)
         |        |FP:  Direction                        |Uplink
         |        |FP:  Transport Channel Type           |DCH (Dedicated Channel)
|1110100- |FP:  Frame CRC                                |'74'H
|-------1 |FP:  Frame Type                               |Control
|00000100 |FP:  Control Frame Type (DCH)                 |UL Synchronisation
|11110000 |FP:  Connection Frame Number                  |240
|***B2*** |FP:  ToA (0,125 ms)                           |198


+---------+--------------------------------------+------------------------------------+
|BITMASK  |ID Name                               |Comment or Value                    |           12
+---------+--------------------------------------+------------------------------------+
|TS 25.322 V3.10.0 (RLC) / 25.321 V3.11.0 (MAC) / 25.435 V3.10.0, 25.427 V3.9.0 (FP) - (2002-03) (RLC/MAC)   FP CTRL DCH
(= FP Control Frame DCH)  |
|FP Control Frame DCH
         |        |FP:  VPI/VCI/CID                      |"14/06/66"
         |        |FP:  Radio Mode                       |FDD (Frequency Division Duplex)
```

UMTS – Signaling & Protocol Analysis

```
         |        |FP:  Direction                        |Downlink
         |        |FP:  Transport Channel Type           |DCH (Dedicated Channel)
|1101111- |FP:  Frame CRC                                |6f
|-------1 |FP:  Frame Type                               |Control
|00000011 |FP:  Control Frame Type (DCH)                 |DL Synchronisation
|11110001 |FP:  Connection Frame Number                  |241


+---------+--------------------------------------+------------------------------------+
|BITMASK  |ID Name                               |Comment or Value                    |           13
+---------+--------------------------------------+------------------------------------+
|TS 25.322 V3.10.0 (RLC) / 25.321 V3.11.0 (MAC) / 25.435 V3.10.0, 25.427 V3.9.0 (FP) - (2002-03) (RLC/MAC)   FP CTRL DCH
(= FP Control Frame DCH)  |
|FP Control Frame DCH
         |        |FP:  VPI/VCI/CID                      |"14/06/66"
         |        |FP:  Radio Mode                       |FDD (Frequency Division Duplex)
         |        |FP:  Direction                        |Uplink
         |        |FP:  Transport Channel Type           |DCH (Dedicated Channel)
|0010010- |FP:  Frame CRC                                |'12'H
|-------1 |FP:  Frame Type                               |Control
|00000100 |FP:  Control Frame Type (DCH)                 |UL Synchronisation
|11110001 |FP:  Connection Frame Number                  |241
|***B2*** |FP:  ToA (0,125 ms)                           |196


+---------+--------------------------------------+------------------------------------+
|BITMASK  |ID Name                               |Comment or Value                    |           14
+---------+--------------------------------------+------------------------------------+
|TS 25.322 V3.7.0 (2001-06) reassembled (RLC reasm.)  RLC UMD (= RLC Unacknowledged Mode Data)  |
|RLC Unacknowledged Mode Data
|***B7*** |FP:  VPI/VCI/CID                              |"0/10/12"
         |        |FP:  Direction                        |Downlink
         |        |FP:  Transport Channel Type           |FACH (Forward Access Channel)
         |        |MAC: Target Channel Type              |CCCH (Common Control Channel)
         |        |MAC: RLC Mode                         |Unacknowledge Mode
```

UMTS – Signaling & Protocol Analysis

```
|**B100** |RLC: Whole Data                               |30 6f 12 4c 40 4e 20 01 2c 02 a0...|
|TS 29.331 CCCH-DL (2001-06) (RRC_CCCH_DL)  rrcConnectionSetup (= rrcConnectionSetup)
|dL-CCCH-Message
|1 message
|2 rrcConnectionSetup
|3 r3
|3.1 rrcConnectionSetup-r3
|3.1.1 initialUE-Identity
|3.1.2 imsi
|***b4***  |3.1.2.1 digit                                |2
|---0111-  |3.1.2.2 digit                                |7
|***b4***  |3.1.2.3 digit                                |2
|---0000-  |3.1.2.4 digit                                |0
|***b4***  |3.1.2.5 digit                                |7
|---0111-  |3.1.2.6 digit                                |7
|***b4***  |3.1.2.7 digit                                |1
|---0000-  |3.1.2.8 digit                                |0
|***b4***  |3.1.2.9 digit                                |0
|---0000-  |3.1.2.10 digit                               |0
|***b4***  |3.1.2.11 digit                               |9
|---0110-  |3.1.2.12 digit                               |6
|***b4***  |3.1.2.13 digit                               |0
|---0001-  |3.1.2.14 digit                               |1
|***b4***  |3.1.2.15 digit                               |5
|---00---  |3.1.2 rrc-TransactionIdentifier              |0
|3.1.3 new-U-RNTI
|**b12***  |3.1.3.1 srnc-Identity                        |'00000000110'B
|**b20***  |3.1.3.2 s-RNTI                               |'00000100110101111011'B
|-----00-  |3.1.4 rrc-StateIndicator                     |cell-DCH
|***b3***  |3.1.5 utran-DRX-CycleLengthCoeff             |3
|3.1.6 srb-InformationSetupList
|3.1.6.1 sRB-InformationSetup
|***b5***  |3.1.6.1.1 rb-Identity                        |1
|3.1.6.1.2 rlc-InfoChoice
|3.1.6.1.3 rlc-Info
```

```
|3.1.6.1.3.1 ul-RLC-Mode
|3.1.6.1.3.2 ul-UM-RLC-Mode
|3.1.6.1.3.2 dl-RLC-Mode
|          |3.1.6.1.3.3 dl-UM-RLC-Mode                   |0
|3.1.6.1.3 rb-MappingInfo
|3.1.6.1.3.1 rB-MappingOption
|3.1.6.1.3.1.1 ul-LogicalChannelMappings
|3.1.6.1.3.1.2 oneLogicalChannel
|3.1.6.1.3.1.2.1 ul-TransportChannelType
|--11111-  |3.1.6.1.3.1.2.2 dch                          |32
|***b4***  |3.1.6.1.3.1.2.2 logicalChannelIdentity       |2
|3.1.6.1.3.1.2.3 rlc-SizeList
|          |3.1.6.1.3.1.2.4 configured                   |0
|-----000  |3.1.6.1.3.1.2.4 mac-LogicalChannelPriority   |1
|3.1.6.1.3.1.2 dl-LogicalChannelMappingList
|3.1.6.1.3.1.2.1 dL-LogicalChannelMapping
|3.1.6.1.3.1.2.1.1 dl-TransportChannelType
|***b5***  |3.1.6.1.3.1.2.1.2 dch                        |32
|-0001---  |3.1.6.1.3.1.2.1.2 logicalChannelIdentity     |2
|3.1.6.2 sRB-InformationSetup
|***b5***  |3.1.6.2.1 rb-Identity                        |2
|3.1.6.2.2 rlc-InfoChoice
|3.1.6.2.3 rlc-Info
|3.1.6.2.3.1 ul-RLC-Mode
|3.1.6.2.3.2 ul-AM-RLC-Mode
|3.1.6.2.3.2.1 transmissionRLC-Discard
|---1010-  |3.1.6.2.3.2.2 noDiscard                      |dat15
|***b4***  |3.1.6.2.3.2.2 transmissionWindowSize         |tw128
|---0010-  |3.1.6.2.3.2.3 timerRST                       |tr150
|***b3***  |3.1.6.2.3.2.4 max-RST                        |rst1
|3.1.6.2.3.2.5 pollingInfo
|100111--  |3.1.6.2.3.2.5.1 timerPoll                    |tp400
|------00  |3.1.6.2.3.2.5.2 poll-SDU                     |sdu1
|1-------  |3.1.6.2.3.2.5.3 lastTransmissionPDU-Poll     |1
|-1------  |3.1.6.2.3.2.5.4 lastRetransmissionPDU-Poll   |1
```

UMTS – Signaling & Protocol Analysis

```
|--010---  |3.1.6.2.3.2.5.5 pollWindow                      |pw70     |
|3.1.6.2.3.2 dl-RLC-Mode                                              |
|3.1.6.2.3.3 dl-AM-RLC-Mode                                           |
|-------1  |3.1.6.2.3.3.1 inSequenceDelivery                |1        |
|0101----  |3.1.6.2.3.3.2 receivingWindowSize               |rw128    |
|3.1.6.2.3.3.3 dl-RLC-StatusInfo                                      |
|***b6***  |3.1.6.2.3.3.3.1 timerStatusProhibit             |tsp120   |
|-----1--  |3.1.6.2.3.3.3.2 missingPDU-Indicator            |1        |
|3.1.6.2.3 rb-MappingInfo                                             |
|3.1.6.2.3.1 rB-MappingOption                                         |
|3.1.6.2.3.1.1 ul-LogicalChannelMappings                              |
|3.1.6.2.3.1.2 oneLogicalChannel                                      |
|3.1.6.2.3.1.2.1 ul-TransportChannelType                              |
|***b5***  |3.1.6.2.3.1.2.2 dch                             |32       |
|----0010  |3.1.6.2.3.1.2.2 logicalChannelIdentity          |3        |
|3.1.6.2.3.1.2.3 rlc-SizeList                                         |
|          |3.1.6.2.3.1.2.4 configured                      |0        |
|--000---  |3.1.6.2.3.1.2.4 mac-LogicalChannelPriority      |1        |
|3.1.6.2.3.1.2 dl-LogicalChannelMappingList                           |
|3.1.6.2.3.1.2.1 dL-LogicalChannelMapping                             |
|3.1.6.2.3.1.2.1.1 dl-TransportChannelType                            |
|-11111--  |3.1.6.2.3.1.2.1.2 dch                           |32       |
|***b4***  |3.1.6.2.3.1.2.1.2 logicalChannelIdentity        |3        |
|3.1.6.3 sRB-InformationSetup                                         |
|---00010  |3.1.6.3.1 rb-Identity                           |3        |
|3.1.6.3.2 rlc-InfoChoice                                             |
|3.1.6.3.3 rlc-Info                                                   |
|3.1.6.3.3.1 ul-RLC-Mode                                              |
|3.1.6.3.3.2 ul-AM-RLC-Mode                                           |
|3.1.6.3.3.2.1 transmissionRLC-Discard                                |
|1010----  |3.1.6.3.3.2.2 noDiscard                         |dat15    |
|----0101  |3.1.6.3.3.2.2 transmissionWindowSize            |tw128    |
|0010----  |3.1.6.3.3.2.3 timerRST                          |tr150    |
|----000-  |3.1.6.3.3.2.4 max-RST                           |rst1     |
|3.1.6.3.3.2.5 pollingInfo                                            |
```

```
|***b6***  |3.1.6.3.3.2.5.1 timerPoll                       |tp400    |
|---00---  |3.1.6.3.3.2.5.2 poll-SDU                        |sdu1     |
|-----1--  |3.1.6.3.3.2.5.3 lastTransmissionPDU-Poll        |1        |
|------1-  |3.1.6.3.3.2.5.4 lastRetransmissionPDU-Poll      |1        |
|***b3***  |3.1.6.3.3.2.5.5 pollWindow                      |pw70     |
|3.1.6.3.3.2 dl-RLC-Mode                                              |
|3.1.6.3.3.3 dl-AM-RLC-Mode                                           |
|----1---  |3.1.6.3.3.3.1 inSequenceDelivery                |1        |
|***b4***  |3.1.6.3.3.3.2 receivingWindowSize               |rw128    |
|3.1.6.3.3.3.3 dl-RLC-StatusInfo                                      |
|***b6***  |3.1.6.3.3.3.3.1 timerStatusProhibit             |tsp120   |
|--1-----  |3.1.6.3.3.3.3.2 missingPDU-Indicator            |1        |
|3.1.6.3.3 rb-MappingInfo                                             |
|3.1.6.3.3.1 rB-MappingOption                                         |
|3.1.6.3.3.1.1 ul-LogicalChannelMappings                              |
|3.1.6.3.3.1.2 oneLogicalChannel                                      |
|3.1.6.3.3.1.2.1 ul-TransportChannelType                              |
|***b5***  |3.1.6.3.3.1.2.2 dch                             |32       |
|-0011---  |3.1.6.3.3.1.2.2 logicalChannelIdentity          |4        |
|3.1.6.3.3.1.2.3 rlc-SizeList                                         |
|          |3.1.6.3.3.1.2.4 configured                      |0        |
|***b3***  |3.1.6.3.3.1.2.4 mac-LogicalChannelPriority      |1        |
|3.1.6.3.3.1.2 dl-LogicalChannelMappingList                           |
|3.1.6.3.3.1.2.1 dL-LogicalChannelMapping                             |
|3.1.6.3.3.1.2.1.1 dl-TransportChannelType                            |
|***b5***  |3.1.6.3.3.1.2.1.2 dch                           |32       |
|---0011-  |3.1.6.3.3.1.2.1.2 logicalChannelIdentity        |4        |
|3.1.6.4 sRB-InformationSetup                                         |
|00011---  |3.1.6.4.1 rb-Identity                           |4        |
|3.1.6.4.2 rlc-InfoChoice                                             |
|3.1.6.4.3 rlc-Info                                                   |
|3.1.6.4.3.1 ul-RLC-Mode                                              |
|3.1.6.4.3.2 ul-AM-RLC-Mode                                           |
|3.1.6.4.3.2.1 transmissionRLC-Discard                                |
|***b4***  |3.1.6.4.3.2.2 noDiscard                         |dat15    |
```

UMTS – Signaling & Protocol Analysis

```
|-0101---  |3.1.6.4.3.2.2 transmissionWindowSize        |tw128
|***b4***  |3.1.6.4.3.2.3 timerRST                      |tr150
|-000----  |3.1.6.4.3.2.4 max-RST                       |rst1
|3.1.6.4.3.2.5 pollingInfo
|--100111  |3.1.6.4.3.2.5.1 timerPoll                   |tp400
|00------  |3.1.6.4.3.2.5.2 poll-SDU                    |sdu1
|--1-----  |3.1.6.4.3.2.5.3 lastTransmissionPDU-Poll    |1
|---1----  |3.1.6.4.3.2.5.4 lastRetransmissionPDU-Poll  |1
|----010-  |3.1.6.4.3.2.5.5 pollWindow                  |pw70
|3.1.6.4.3.2 dl-RLC-Mode
|3.1.6.4.3.3 dl-AM-RLC-Mode
|-1------  |3.1.6.4.3.3.1 inSequenceDelivery            |1
|--0101--  |3.1.6.4.3.3.2 receivingWindowSize           |rw128
|3.1.6.4.3.3.3 dl-RLC-StatusInfo
|-001011-  |3.1.6.4.3.3.3.1 timerStatusProhibit         |tsp120
|-------1  |3.1.6.4.3.3.3.2 missingPDU-Indicator        |1
|3.1.6.4.3 rb-MappingInfo
|3.1.6.4.3.1 rB-MappingOption
|3.1.6.4.3.1.1 ul-LogicalChannelMappings
|3.1.6.4.3.1.2 oneLogicalChannel
|3.1.6.4.3.1.2.1 ul-TransportChannelType
|-11111--  |3.1.6.4.3.1.2.2 dch                         |32
|***b4***  |3.1.6.4.3.1.2.2 logicalChannelIdentity      |5
|3.1.6.4.3.1.2.3 rlc-SizeList
|          |3.1.6.4.3.1.2.4 configured                  |0
|----000-  |3.1.6.4.3.1.2.4 mac-LogicalChannelPriority  |1
|3.1.6.4.3.1.2 dl-LogicalChannelMappingList
|3.1.6.4.3.1.2.1 dL-LogicalChannelMapping
|3.1.6.4.3.1.2.1.1 dl-TransportChannelType
|---11111  |3.1.6.4.3.1.2.1.2 dch                       |32
|0100----  |3.1.6.4.3.1.2.1.2 logicalChannelIdentity    |5
|3.1.7 ul-CommonTransChInfo
|3.1.7.1 modeSpecificInfo
|3.1.7.2 fdd
|3.1.7.2.1 ul-TFCS
```

UMTS – Signaling & Protocol Analysis

```
|3.1.7.2.2 normalTFCI-Signalling
|3.1.7.2.3 complete
|3.1.7.2.3.1 ctfcSize
|3.1.7.2.3.2 ctfc2Bit
|3.1.7.2.3.2.1 sequence
|-00-----  |3.1.7.2.3.2.1.1 ctfc2                       |0
|3.1.7.2.3.2.1.2 powerOffsetInformation
|3.1.7.2.3.2.1.2.1 gainFactorInformation
|3.1.7.2.3.2.1.2.2 signalledGainFactors
|3.1.7.2.3.2.1.2.2.1 modeSpecificInfo
|3.1.7.2.3.2.1.2.2.2 fdd
|***b4***  |3.1.7.2.3.2.1.2.2.2.1 gainFactorBetaC       |15
|---1010-  |3.1.7.2.3.2.1.2.2.2.2 gainFactorBetaD       |10
|***b2***  |3.1.7.2.3.2.1.2.2.2.3 referenceTFC-ID       |0
|3.1.7.2.3.2.2 sequence
|--01----  |3.1.7.2.3.2.2.1 ctfc2                       |1
|3.1.7.2.3.2.2.2 powerOffsetInformation
|3.1.7.2.3.2.2.2.1 gainFactorInformation
|3.1.7.2.3.2.2.2.2 signalledGainFactors
|3.1.7.2.3.2.2.2.2.1 modeSpecificInfo
|3.1.7.2.3.2.2.2.2.2 fdd
|1111----  |3.1.7.2.3.2.2.2.2.2.1 gainFactorBetaC       |15
|----1111  |3.1.7.2.3.2.2.2.2.2.2 gainFactorBetaD       |15
|00------  |3.1.7.2.3.2.2.2.2.2.3 referenceTFC-ID       |0
|3.1.8 ul-AddReconfTransChInfoList
|3.1.8.1 uL-AddReconfTransChInformation
|-------0  |3.1.8.1.1 ul-TransportChannelType           |dch
|11111---  |3.1.8.1.2 transportChannelIdentity          |32
|3.1.8.1.3 transportFormatSet
|3.1.8.1.4 dedicatedTransChTFS
|3.1.8.1.4.1 tti
|3.1.8.1.4.2 tti40
|3.1.8.1.4.2.1 dedicatedDynamicTF-Info
|3.1.8.1.4.2.1.1 rlc-Size
|3.1.8.1.4.2.1.2 octetModeType1
```

UMTS – Signaling & Protocol Analysis

```
|-10000-- |3.1.8.1.4.2.1.3 sizeType1                    |16
|3.1.8.1.4.2.1.2 numberOfTbSizeList
|3.1.8.1.4.2.1.2.1 numberOfTransportBlocks
|         |3.1.8.1.4.2.1.2.2 zero                       |0
|3.1.8.1.4.2.1.3 logicalChannelList
|         |3.1.8.1.4.2.1.4 allSizes                     |0
|3.1.8.1.4.2.2 dedicatedDynamicTF-Info
|3.1.8.1.4.2.2.1 rlc-Size
|3.1.8.1.4.2.2.2 octetModeType1
|--10000- |3.1.8.1.4.2.2.3 sizeType1                    |16
|3.1.8.1.4.2.2.2 numberOfTbSizeList
|3.1.8.1.4.2.2.2.1 numberOfTransportBlocks
|         |3.1.8.1.4.2.2.2.2 one                       |0
|3.1.8.1.4.2.2.3 logicalChannelList
|         |3.1.8.1.4.2.2.4 allSizes                     |0
|3.1.8.1.4.2 semistaticTF-Information
|3.1.8.1.4.2.1 channelCodingType
|--1----- |3.1.8.1.4.2.2 convolutional                  |third
|***b8*** |3.1.8.1.4.2.2 rateMatchingAttribute          |160
|---011-- |3.1.8.1.4.2.3 crc-Size                       |crc16
|3.1.9 dl-CommonTransChInfo
|3.1.9.1 modeSpecificInfo
|3.1.9.2 fdd
|3.1.9.2.1 dl-Parameters
|3.1.9.2.2 dl-DCH-TFCS
|3.1.9.2.3 normalTFCI-Signalling
|3.1.9.2.4 complete
|3.1.9.2.4.1 ctfcSize
|3.1.9.2.4.2 ctfc2Bit
|3.1.9.2.4.2.1 sequence
|---00--- |3.1.9.2.4.2.1.1 ctfc2                        |0
|3.1.9.2.4.2.2 sequence
|------01 |3.1.9.2.4.2.2.1 ctfc2                        |1
|3.1.10 dl-AddReconfTransChInfoList
|3.1.10.1 dL-AddReconfTransChInformation
```

```
|-------0 |3.1.10.1.1 dl-TransportChannelType           |dch
|11111--- |3.1.10.1.2 dl-transportChannelIdentity       |32
|3.1.10.1.3 tfs-SignallingMode
|3.1.10.1.4 sameAsULTrCH
|------0- |3.1.10.1.4.1 ul-TransportChannelType         |dch
|***b5*** |3.1.10.1.4.2 ul-TransportChannelIdentity     |32
|3.1.10.1.4 dch-QualityTarget
|***b6*** |3.1.10.1.4.1 bler-QualityValue               |-20
|***b7*** |3.1.11 maxAllowedUL-TX-Power                 |33
|3.1.12 ul-ChannelRequirement
|3.1.13 ul-DPCH-Info
|3.1.13.1 ul-DPCH-PowerControlInfo
|3.1.13.2 fdd
|***b7*** |3.1.13.2.1 dpcch-PowerOffset                 |-33
|---000-- |3.1.13.2.2 pc-Preamble                       |0
|***b3*** |3.1.13.2.3 sRB-delay                         |0
|3.1.13.2.4 powerControlAlgorithm
|--1----- |3.1.13.2.5 algorithm1                        |1
|3.1.13.2 modeSpecificInfo
|3.1.13.3 fdd
|-------1- |3.1.13.3.1 scramblingCodeType               |longSC
|**b24*** |3.1.13.3.2 scramblingCode                    |1068411
|         |3.1.13.3.3 numberOfDPDCH                     |1
|***b3*** |3.1.13.3.4 spreadingFactor                   |sf256
|--1----- |3.1.13.3.5 tfci-Existence                    |1
|---1111- |3.1.13.3.6 puncturingLimit                   |pl1
|3.1.13 dl-CommonInformation
|3.1.13.1 dl-DPCH-InfoCommon
|3.1.13.1.1 cfnHandling
|3.1.13.1.2 initialise
|3.1.13.1.2 modeSpecificInfo
|3.1.13.1.3 fdd
|3.1.13.1.3.1 dl-DPCH-PowerControlInfo
|3.1.13.1.3.1.1 modeSpecificInfo
|3.1.13.1.3.1.2 fdd
```

UMTS – Signaling & Protocol Analysis

```
|------0- |3.1.13.1.3.1.2.1 dpc-Mode                |singleTPC
|***b5*** |3.1.13.1.3.2 powerOffsetPilot-pdpdch     |0
|3.1.13.1.3.3 spreadingFactorAndPilot
|***b2*** |3.1.13.1.3.4 sfd256                      |pb4
|-0------ |3.1.13.1.3.4 positionFixedOrFlexible     |fixed
|--0----- |3.1.13.1.3.5 tfci-Existence              |0
|3.1.13.2 modeSpecificInfo
|3.1.13.3 fdd
|**b10*** |3.1.13.3.1 defaultDPCH-OffsetValue       |0
|--00---- |3.1.13.3.2 tx-DiversityMode              |noDiversity
|3.1.14 dl-InformationPerRL-List
|3.1.14.1 dL-InformationPerRL
|3.1.14.1.1 modeSpecificInfo
|3.1.14.1.2 fdd
|3.1.14.1.2.1 primaryCPICH-Info
|***b9*** |3.1.14.1.2.1.1 primaryScramblingCode     |1
|3.1.14.1.2 dl-DPCH-InfoPerRL
|3.1.14.1.3 fdd
|-0------ |3.1.14.1.3.1 pCPICH-UsageForChannelEst   |mayBeUsed
|***b8*** |3.1.14.1.3.2 dpch-FrameOffset            |0
|3.1.14.1.3.3 dl-ChannelisationCodeList
|3.1.14.1.3.3.1 dL-ChannelisationCode
|3.1.14.1.3.3.1.1 sf-AndCodeNumber
|***b8*** |3.1.14.1.3.3.1.2 sf256                   |4
|--000--- |3.1.14.1.3.4 tpc-CombinationIndex        |0


+---------+--------------------------------------+-----------------------------------+
|BITMASK  |ID Name                               |Comment or Value                   |
+---------+--------------------------------------+-----------------------------------+
|TS 25.322 V3.10.0 (2002-03) reassembled (RLC reasm.)  UM DATA FACH (= Unacknowledged Mode Data FACH)
|Unacknowledged Mode Data FACH
|         |FP:  VPI/VCI/CID                      |"0/10/12"
|         |FP:  Direction                        |Downlink
```

15

UMTS – Signaling & Protocol Analysis

```
|         |FP:  Transport Channel Type           |FACH (Forward Access Channel)
|         |MAC: Target Channel Type              |CCCH (Common Control Channel)
|         |MAC: RLC Mode                         |Unacknowledge Mode
|**B104** |RLC: Whole Data                       |30 e7 25 c0 00 5d 44 c4 76 00 10...
|TS 29.331 CCCH-DL (2002-03) (RRC_CCCH_DL)  rrcConnectionSetup (= rrcConnectionSetup)
|dL-CCCH-Message
|1 message
|1.1 rrcConnectionSetup
|1.1.1 r3
|1.1.1.1 rrcConnectionSetup-r3
|1.1.1.1.1 initialUE-Identity
|1.1.1.1.1.1 tmsi-and-LAI
|**b32*** |1.1.1.1.1.1.1 tmsi                    |'00101110000000000000001011101010'B
|1.1.1.1.1.2 lai
|1.1.1.1.1.2.1 plmn-Identity
|1.1.1.1.1.2.1.1 mcc
|---0010- |1.1.1.1.1.2.1.1.1 digit               |2
|***b4*** |1.1.1.1.1.2.1.1.2 digit               |6
|---0010- |1.1.1.1.1.2.1.1.3 digit               |2
|1.1.1.1.1.2.1.2 mnc
|0111---- |1.1.1.1.1.2.1.2.1 digit               |7
|----0110 |1.1.1.1.1.2.1.2.2 digit               |6
|***B2*** |1.1.1.1.1.2.2 lac                     |(3)
|00------ |1.1.1.1.2 rrc-TransactionIdentifier   |0
|1.1.1.1.3 new-U-RNTI
|**b12*** |1.1.1.1.3.1 srnc-Identity             |'000000100010'B
|**b20*** |1.1.1.1.3.2 s-RNTI                    |'00001000101011001 0'B
|--00---- |1.1.1.1.4 rrc-StateIndicator          |cell-DCH
|----000- |1.1.1.1.5 utran-DRX-CycleLengthCoeff  |3
|1.1.1.1.6 capabilityUpdateRequirement
|1------- |1.1.1.1.6.1 ue-RadioCapabilityFDDUpdateRequ..|1
|-0------ |1.1.1.1.6.2 ue-RadioCapabilityTDDUpdateRequ..|0
|1.1.1.1.6.3 systemSpecificCapUpdateReqList
|         |1.1.1.1.6.3.1 systemSpecificCapUpdateReq|gsm
|1.1.1.1.7 srb-InformationSetupList
```

UMTS – Signaling & Protocol Analysis

```
|1.1.1.1.7.1 sRB-InformationSetup
|00000---  |1.1.1.1.7.1.1 rb-Identity                   |1
|1.1.1.1.7.1.2 rlc-InfoChoice
|1.1.1.1.7.1.2.1 rlc-Info
|1.1.1.1.7.1.2.1.1 ul-RLC-Mode
|1.1.1.1.7.1.2.1.1.1 ul-UM-RLC-Mode
|1.1.1.1.7.1.2.1.2 dl-RLC-Mode
|          |1.1.1.1.7.1.2.1.2.1 dl-UM-RLC-Mode          |0
|1.1.1.1.7.1.3 rb-MappingInfo
|1.1.1.1.7.1.3.1 rB-MappingOption
|1.1.1.1.7.1.3.1.1 ul-LogicalChannelMappings
|1.1.1.1.7.1.3.1.1.1 oneLogicalChannel
|1.1.1.1.7.1.3.1.1.1.1 ul-TransportChannelType
|***b5***  |1.1.1.1.7.1.3.1.1.1.1.1 dch                 |32
|---0001-  |1.1.1.1.7.1.3.1.1.1.2 logicalChannelIdentity|2
|1.1.1.1.7.1.3.1.1.1.3 rlc-SizeList
|          |1.1.1.1.7.1.3.1.1.1.3.1 configured         |0
|-000----  |1.1.1.1.7.1.3.1.1.1.4 mac-LogicalChannelPri..|1
|1.1.1.1.7.1.3.1.2 dl-LogicalChannelMappingList
|1.1.1.1.7.1.3.1.2.1 dL-LogicalChannelMapping
|1.1.1.1.7.1.3.1.2.1.1 dl-TransportChannelType
|11111---  |1.1.1.1.7.1.3.1.2.1.1.1 dch                 |32
|***b4***  |1.1.1.1.7.1.3.1.2.1.2 logicalChannelIdentity|2
|1.1.1.1.7.1.3.2 rB-MappingOption
|1.1.1.1.7.1.3.2.1 ul-LogicalChannelMappings
|1.1.1.1.7.1.3.2.1.1 oneLogicalChannel
|1.1.1.1.7.1.3.2.1.1.1 ul-TransportChannelType
|          |1.1.1.1.7.1.3.2.1.1.1.1 rach               |0
|***b4***  |1.1.1.1.7.1.3.2.1.1.2 logicalChannelIdentity|2
|1.1.1.1.7.1.3.2.1.1.3 rlc-SizeList
|1.1.1.1.7.1.3.2.1.1.3.1 explicitList
|1.1.1.1.7.1.3.2.1.1.3.1.1 rLC-SizeInfo
|--00000-  |1.1.1.1.7.1.3.2.1.1.3.1.1.1 rlc-SizeIndex   |1
|***b3***  |1.1.1.1.7.1.3.2.1.1.4 mac-LogicalChannelPri..|1
|1.1.1.1.7.1.3.2.2 dl-LogicalChannelMappingList
```

UMTS – Signaling & Protocol Analysis

```
|1.1.1.1.7.1.3.2.2.1 dL-LogicalChannelMapping
|1.1.1.1.7.1.3.2.2.1.1 dl-TransportChannelType
|          |1.1.1.1.7.1.3.2.2.1.1.1 fach               |0
|***b4***  |1.1.1.1.7.1.3.2.2.1.2 logicalChannelIdentity|2
|1.1.1.1.7.2 sRB-InformationSetup
|---00001  |1.1.1.1.7.2.1 rb-Identity                   |2
|1.1.1.1.7.2.2 rlc-InfoChoice
|1.1.1.1.7.2.2.1 rlc-Info
|1.1.1.1.7.2.2.1.1 ul-RLC-Mode
|1.1.1.1.7.2.2.1.1.1 ul-AM-RLC-Mode
|1.1.1.1.7.2.2.1.1.1.1 transmissionRLC-Discard
|1010----  |1.1.1.1.7.2.2.1.1.1.1.1 noDiscard           |dat15
|----0101  |1.1.1.1.7.2.2.1.1.1.2 transmissionWindowSize|tw128
|0010----  |1.1.1.1.7.2.2.1.1.1.3 timerRST              |tr150
|----000-  |1.1.1.1.7.2.2.1.1.1.4 max-RST               |rst1
|1.1.1.1.7.2.2.1.1.1.5 pollingInfo
|***b6***  |1.1.1.1.7.2.2.1.1.1.5.1 timerPoll           |tp400
|---00---  |1.1.1.1.7.2.2.1.1.1.5.2 poll-SDU            |sdu1
|-----1--  |1.1.1.1.7.2.2.1.1.1.5.3 lastTransmissionPDU..|1
|------1-  |1.1.1.1.7.2.2.1.1.1.5.4 lastRetransmissionP..|1
|***b3***  |1.1.1.1.7.2.2.1.1.1.5.5 pollWindow          |pw70
|1.1.1.1.7.2.2.1.2 dl-RLC-Mode
|1.1.1.1.7.2.2.1.2.1 dl-AM-RLC-Mode
|----1---  |1.1.1.1.7.2.2.1.2.1.1 inSequenceDelivery    |1
|***b4***  |1.1.1.1.7.2.2.1.2.1.2 receivingWindowSize   |rw128
|1.1.1.1.7.2.2.1.2.1.3 dl-RLC-StatusInfo
|***b6***  |1.1.1.1.7.2.2.1.2.1.3.1 timerStatusProhibit |tsp120
|--1-----  |1.1.1.1.7.2.2.1.2.1.3.2 missingPDU-Indicator|1
|1.1.1.1.7.2.3 rb-MappingInfo
|1.1.1.1.7.2.3.1 rB-MappingOption
|1.1.1.1.7.2.3.1.1 ul-LogicalChannelMappings
|1.1.1.1.7.2.3.1.1.1 oneLogicalChannel
|1.1.1.1.7.2.3.1.1.1.1 ul-TransportChannelType
|***b5***  |1.1.1.1.7.2.3.1.1.1.1.1 dch                 |32
|-0010---  |1.1.1.1.7.2.3.1.1.1.2 logicalChannelIdentity|3
```

UMTS – Signaling & Protocol Analysis

```
         |1.1.1.1.7.2.3.1.1.1.3 rlc-SizeList
         |    |1.1.1.1.7.2.3.1.1.1.3.1 configured          |0
|***b3***|1.1.1.1.7.2.3.1.1.1.4 mac-LogicalChannelPri..|1
|1.1.1.1.7.2.3.1.2 dl-LogicalChannelMappingList
|1.1.1.1.7.2.3.1.2.1 dL-LogicalChannelMapping
|1.1.1.1.7.2.3.1.2.1.1 dl-TransportChannelType
|***b5***|1.1.1.1.7.2.3.1.2.1.1.1 dch                 |32
|---0010-|1.1.1.1.7.2.3.1.2.1.2 logicalChannelIdentity|3
|1.1.1.1.7.2.3.2 rB-MappingOption
|1.1.1.1.7.2.3.2.1 ul-LogicalChannelMappings
|1.1.1.1.7.2.3.2.1.1 oneLogicalChannel
|1.1.1.1.7.2.3.2.1.1.1 ul-TransportChannelType
|        |1.1.1.1.7.2.3.2.1.1.1.1 rach                |0
|***b4***|1.1.1.1.7.2.3.2.1.1.2 logicalChannelIdentity|3
|1.1.1.1.7.2.3.2.1.1.3 rlc-SizeList
|1.1.1.1.7.2.3.2.1.1.3.1 explicitList
|1.1.1.1.7.2.3.2.1.1.3.1.1 rLC-SizeInfo
|00000---|1.1.1.1.7.2.3.2.1.1.3.1.1.1 rlc-SizeIndex   |1
|-----000|1.1.1.1.7.2.3.2.1.1.4 mac-LogicalChannelPri..|1
|1.1.1.1.7.2.3.2.2 dl-LogicalChannelMappingList
|1.1.1.1.7.2.3.2.2.1 dL-LogicalChannelMapping
|1.1.1.1.7.2.3.2.2.1.1 dl-TransportChannelType
|        |1.1.1.1.7.2.3.2.2.1.1.1 fach                |0
|----0010|1.1.1.1.7.2.3.2.2.1.2 logicalChannelIdentity|3
|1.1.1.1.7.3 sRB-InformationSetup
|-00010--|1.1.1.1.7.3.1 rb-Identity                   |3
|1.1.1.1.7.3.2 rlc-InfoChoice
|***b5***|1.1.1.1.7.3.2.1 same-as-RB                  |2
|1.1.1.1.7.3.3 rb-MappingInfo
|1.1.1.1.7.3.3.1 rB-MappingOption
|1.1.1.1.7.3.3.1.1 ul-LogicalChannelMappings
|1.1.1.1.7.3.3.1.1.1 oneLogicalChannel
|1.1.1.1.7.3.3.1.1.1.1 ul-TransportChannelType
|***b5***|1.1.1.1.7.3.3.1.1.1.1.1 dch                 |32
|--0011--|1.1.1.1.7.3.3.1.1.2 logicalChannelIdentity  |4
```

UMTS – Signaling & Protocol Analysis

```
|1.1.1.1.7.3.3.1.1.1.3 rlc-SizeList
|        |1.1.1.1.7.3.3.1.1.1.3.1 configured          |0
|000-----|1.1.1.1.7.3.3.1.1.1.4 mac-LogicalChannelPri..|1
|1.1.1.1.7.3.3.1.2 dl-LogicalChannelMappingList
|1.1.1.1.7.3.3.1.2.1 dL-LogicalChannelMapping
|1.1.1.1.7.3.3.1.2.1.1 dl-TransportChannelType
|***b5***|1.1.1.1.7.3.3.1.2.1.1.1 dch                 |32
|----0011|1.1.1.1.7.3.3.1.2.1.2 logicalChannelIdentity|4
|1.1.1.1.7.3.3.2 rB-MappingOption
|1.1.1.1.7.3.3.2.1 ul-LogicalChannelMappings
|1.1.1.1.7.3.3.2.1.1 oneLogicalChannel
|1.1.1.1.7.3.3.2.1.1.1 ul-TransportChannelType
|        |1.1.1.1.7.3.3.2.1.1.1.1 rach                |0
|***b4***|1.1.1.1.7.3.3.2.1.1.2 logicalChannelIdentity|4
|1.1.1.1.7.3.3.2.1.1.3 rlc-SizeList
|1.1.1.1.7.3.3.2.1.1.3.1 explicitList
|1.1.1.1.7.3.3.2.1.1.3.1.1 rLC-SizeInfo
|-00000--|1.1.1.1.7.3.3.2.1.1.3.1.1.1 rlc-SizeIndex   |1
|***b3***|1.1.1.1.7.3.3.2.1.1.4 mac-LogicalChannelPri..|1
|1.1.1.1.7.3.3.2.2 dl-LogicalChannelMappingList
|1.1.1.1.7.3.3.2.2.1 dL-LogicalChannelMapping
|1.1.1.1.7.3.3.2.2.1.1 dl-TransportChannelType
|        |1.1.1.1.7.3.3.2.2.1.1.1 fach                |0
|***b4***|1.1.1.1.7.3.3.2.2.1.2 logicalChannelIdentity|4
|1.1.1.1.7.4 sRB-InformationSetup
|--00011-|1.1.1.1.7.4.1 rb-Identity                   |4
|1.1.1.1.7.4.2 rlc-InfoChoice
|00001---|1.1.1.1.7.4.2.1 same-as-RB                  |2
|1.1.1.1.7.4.3 rb-MappingInfo
|1.1.1.1.7.4.3.1 rB-MappingOption
|1.1.1.1.7.4.3.1.1 ul-LogicalChannelMappings
|1.1.1.1.7.4.3.1.1.1 oneLogicalChannel
|1.1.1.1.7.4.3.1.1.1.1 ul-TransportChannelType
|***b5***|1.1.1.1.7.4.3.1.1.1.1.1 dch                 |32
|---0100-|1.1.1.1.7.4.3.1.1.2 logicalChannelIdentity  |5
```

```
|1.1.1.1.7.4.3.1.1.1.3 rlc-SizeList
|         |1.1.1.1.7.4.3.1.1.1.3.1 configured      |0
|-000---- |1.1.1.1.7.4.3.1.1.1.4 mac-LogicalChannelPri..|1
|1.1.1.1.7.4.3.1.2 dl-LogicalChannelMappingList
|1.1.1.1.7.4.3.1.2.1 dL-LogicalChannelMapping
|1.1.1.1.7.4.3.1.2.1.1 dl-TransportChannelType
|11111--- |1.1.1.1.7.4.3.1.2.1.1.1 dch              |32
|***b4*** |1.1.1.1.7.4.3.1.2.1.2 logicalChannelIdentity |5
|1.1.1.1.7.4.3.2 rB-MappingOption
|1.1.1.1.7.4.3.2.1 ul-LogicalChannelMappings
|1.1.1.1.7.4.3.2.1.1 oneLogicalChannel
|1.1.1.1.7.4.3.2.1.1.1 ul-TransportChannelType
|         |1.1.1.1.7.4.3.2.1.1.1.1 rach            |0
|***b4*** |1.1.1.1.7.4.3.2.1.1.2 logicalChannelIdentity |5
|1.1.1.1.7.4.3.2.1.1.3 rlc-SizeList
|1.1.1.1.7.4.3.2.1.1.3.1 explicitList
|1.1.1.1.7.4.3.2.1.1.3.1.1 rLC-SizeInfo
|--00000- |1.1.1.1.7.4.3.2.1.1.3.1.1.1 rlc-SizeIndex |1
|***b3*** |1.1.1.1.7.4.3.2.1.1.4 mac-LogicalChannelPri..|1
|1.1.1.1.7.4.3.2.2 dl-LogicalChannelMappingList
|1.1.1.1.7.4.3.2.2.1 dL-LogicalChannelMapping
|1.1.1.1.7.4.3.2.2.1.1 dl-TransportChannelType
|         |1.1.1.1.7.4.3.2.2.1.1.1 fach            |0
|***b4*** |1.1.1.1.7.4.3.2.2.1.2 logicalChannelIdentity |5
|1.1.1.1.8 ul-CommonTransChInfo
|1.1.1.1.8.1 modeSpecificInfo
|1.1.1.1.8.1.1 fdd
|1.1.1.1.8.1.1.1 ul-TFCS
|1.1.1.1.8.1.1.1.1 normalTFCI-Signalling
|1.1.1.1.8.1.1.1.1.1 complete
|1.1.1.1.8.1.1.1.1.1.1 ctfcSize
|1.1.1.1.8.1.1.1.1.1.1.1 ctfc2Bit
|1.1.1.1.8.1.1.1.1.1.1.1.1 sequence
|***b2*** |1.1.1.1.8.1.1.1.1.1.1.1.1.1 ctfc2        |0
|1.1.1.1.8.1.1.1.1.1.1.1.1.2 powerOffsetInformation
```

```
|1.1.1.1.8.1.1.1.1.1.1.1.1.2.1 gainFactorInformation
|1.1.1.1.8.1.1.1.1.1.1.1.1.2.1.1 signalledGainFactors
|1.1.1.1.8.1.1.1.1.1.1.1.1.2.1.1.1 modeSpecificInfo
|1.1.1.1.8.1.1.1.1.1.1.1.1.2.1.1.1.1 fdd
|***b4*** |1.1.1.1.8.1.1.1.1.1.1.1.1.2.1.1.1.1.1 gainF..|15
|-1010--- |1.1.1.1.8.1.1.1.1.1.1.1.1.2.1.1.2 gainFacto..|10
|-----00- |1.1.1.1.8.1.1.1.1.1.1.1.1.2.1.1.3 reference..|0
|1.1.1.1.8.1.1.1.1.1.1.1.2 sequence
|01------ |1.1.1.1.8.1.1.1.1.1.1.2.1 ctfc2             |1
|1.1.1.1.8.1.1.1.1.1.1.2.2 powerOffsetInformation
|1.1.1.1.8.1.1.1.1.1.1.2.2.1 gainFactorInformation
|1.1.1.1.8.1.1.1.1.1.1.2.2.1.1 signalledGainFactors
|1.1.1.1.8.1.1.1.1.1.1.2.2.1.1.1 modeSpecificInfo
|1.1.1.1.8.1.1.1.1.1.1.2.2.1.1.1.1 fdd
|***b4*** |1.1.1.1.8.1.1.1.1.1.1.2.2.1.1.1.1.1 gainF..|15
|--1111-- |1.1.1.1.8.1.1.1.1.1.1.2.2.1.1.2 gainFacto..|15
|------00 |1.1.1.1.8.1.1.1.1.1.1.2.2.1.1.3 reference..|0
|1.1.1.1.9 ul-AddReconfTransChInfoList
|1.1.1.1.9.1 uL-AddReconfTransChInformation
|-----0-- |1.1.1.1.9.1.1 ul-TransportChannelType      |dch
|***b5*** |1.1.1.1.9.1.2 transportChannelIdentity     |32
|1.1.1.1.9.1.3 transportFormatSet
|1.1.1.1.9.1.3.1 dedicatedTransChTFS
|1.1.1.1.9.1.3.1.1 tti
|1.1.1.1.9.1.3.1.1.1 tti40
|1.1.1.1.9.1.3.1.1.1.1 dedicatedDynamicTF-Info
|1.1.1.1.9.1.3.1.1.1.1.1 rlc-Size
|1.1.1.1.9.1.3.1.1.1.1.1.1 octetModeType1
|***b5*** |1.1.1.1.9.1.3.1.1.1.1.1.1.1 sizeType1      |16
|1.1.1.1.9.1.3.1.1.1.1.2 numberOfTbSizeList
|1.1.1.1.9.1.3.1.1.1.1.2.1 numberOfTransportBlocks
|         |1.1.1.1.9.1.3.1.1.1.1.2.1.1 zero         |0
|1.1.1.1.9.1.3.1.1.1.1.3 logicalChannelList
|         |1.1.1.1.9.1.3.1.1.1.1.3.1 allSizes       |0
|1.1.1.1.9.1.3.1.1.1.2 dedicatedDynamicTF-Info
```

UMTS – Signaling & Protocol Analysis

```
|1.1.1.1.9.1.3.1.1.1.2.1 rlc-Size                                       |
|1.1.1.1.9.1.3.1.1.1.2.1.1 octetModeType1                               |
|10000---  |1.1.1.1.9.1.3.1.1.1.2.1.1.1 sizeType1         |16           |
|1.1.1.1.9.1.3.1.1.1.2.2 numberOfTbSizeList                             |
|1.1.1.1.9.1.3.1.1.1.2.2.1 numberOfTransportBlocks                      |
|          |1.1.1.1.9.1.3.1.1.1.2.2.1.1 one              |0            |
|1.1.1.1.9.1.3.1.1.1.2.3 logicalChannelList                             |
|          |1.1.1.1.9.1.3.1.1.1.2.3.1 allSizes           |0            |
|1.1.1.1.9.1.3.1.2 semistaticTF-Information                             |
|1.1.1.1.9.1.3.1.2.1 channelCodingType                                  |
|1-------  |1.1.1.1.9.1.3.1.2.1.1 convolutional          |third        |
|***b8***  |1.1.1.1.9.1.3.1.2.2 rateMatchingAttribute    |185          |
|-011----  |1.1.1.1.9.1.3.1.2.3 crc-Size                 |crc16        |
|1.1.1.1.10 dl-CommonTransChInfo                                        |
|1.1.1.1.10.1 modeSpecificInfo                                          |
|1.1.1.1.10.1.1 fdd                                                     |
|1.1.1.1.10.1.1.1 dl-Parameters                                         |
|1.1.1.1.10.1.1.1.1 dl-DCH-TFCS                                         |
|1.1.1.1.10.1.1.1.1.1 normalTFCI-Signalling                             |
|1.1.1.1.10.1.1.1.1.1.1 complete                                        |
|1.1.1.1.10.1.1.1.1.1.1.1 ctfcSize                                      |
|1.1.1.1.10.1.1.1.1.1.1.1.1 ctfc2Bit                                    |
|1.1.1.1.10.1.1.1.1.1.1.1.1.1 sequence                                  |
|-00-----  |1.1.1.1.10.1.1.1.1.1.1.1.1.1.1 ctfc2          |0            |
|1.1.1.1.10.1.1.1.1.1.1.1.1.2 sequence                                  |
|----01--  |1.1.1.1.10.1.1.1.1.1.1.1.1.2.1 ctfc2          |1            |
|1.1.1.1.11 dl-AddReconfTransChInfoList                                 |
|1.1.1.1.11.1 dL-AddReconfTransChInformation                            |
|-----0--  |1.1.1.1.11.1.1 dl-TransportChannelType        |dch          |
|***b5***  |1.1.1.1.11.1.2 dl-transportChannelIdentity   |32           |
|1.1.1.1.11.1.3 tfs-SignallingMode                                      |
|1.1.1.1.11.1.3.1 sameAsULTrCH                                          |
|----0---  |1.1.1.1.11.1.3.1.1 ul-TransportChannelType    |dch          |
|***b5***  |1.1.1.1.11.1.3.1.2 ul-TransportChannelIdent...|32           |
|1.1.1.1.11.1.4 dch-QualityTarget                                       |
```

UMTS – Signaling & Protocol Analysis

```
|--101011  |1.1.1.1.11.1.4.1 bler-QualityValue            |-20          |
|1.1.1.1.12 ul-ChannelRequirement                                       |
|1.1.1.1.12.1 ul-DPCH-Info                                              |
|1.1.1.1.12.1.1 ul-DPCH-PowerControlInfo                                |
|1.1.1.1.12.1.1.1 fdd                                                   |
|***b7***  |1.1.1.1.12.1.1.1.1 dpcch-PowerOffset          |-33          |
|--000---  |1.1.1.1.12.1.1.1.2 pc-Preamble                |0            |
|-----000  |1.1.1.1.12.1.1.1.3 sRB-delay                  |0            |
|1.1.1.1.12.1.1.1.4 powerControlAlgorithm                               |
|-0------  |1.1.1.1.12.1.1.1.4.1 algorithm1               |0            |
|1.1.1.1.12.1.2 modeSpecificInfo                                        |
|1.1.1.1.12.1.2.1 fdd                                                   |
|-----1--  |1.1.1.1.12.1.2.1.1 scramblingCodeType         |longSC       |
|**b24***  |1.1.1.1.12.1.2.1.2 scramblingCode             |1084104      |
|          |1.1.1.1.12.1.2.1.3 numberOfDPDCH              |1            |
|***b3***  |1.1.1.1.12.1.2.1.4 spreadingFactor            |sf256        |
|-1------  |1.1.1.1.12.1.2.1.5 tfci-Existence             |1            |
|--1111--  |1.1.1.1.12.1.2.1.6 puncturingLimit            |pl1          |
|1.1.1.1.13 dl-CommonInformation                                        |
|1.1.1.1.13.1 dl-DPCH-InfoCommon                                        |
|1.1.1.1.13.1.1 cfnHandling                                             |
|1.1.1.1.13.1.1.1 initialise                                            |
|1.1.1.1.13.1.2 modeSpecificInfo                                        |
|1.1.1.1.13.1.2.1 fdd                                                   |
|1.1.1.1.13.1.2.1.1 dl-DPCH-PowerControlInfo                            |
|1.1.1.1.13.1.2.1.1.1 modeSpecificInfo                                  |
|1.1.1.1.13.1.2.1.1.1.1 fdd                                             |
|-----0--  |1.1.1.1.13.1.2.1.1.1.1.1 dpc-Mode             |singleTPC    |
|***b5***  |1.1.1.1.13.1.2.1.2 powerOffsetPilot-pdpdch    |0            |
|1.1.1.1.13.1.2.1.3 spreadingFactorAndPilot                             |
|------01  |1.1.1.1.13.1.2.1.3.1 sfd256                   |pb4          |
|0-------  |1.1.1.1.13.1.2.1.4 positionFixedOrFlexible    |fixed        |
|-0------  |1.1.1.1.13.1.2.1.5 tfci-Existence             |0            |
|1.1.1.1.13.2 modeSpecificInfo                                          |
|1.1.1.1.13.2.1 fdd                                                     |
```

UMTS – Signaling & Protocol Analysis

```
|**b10***  |1.1.1.1.13.2.1.1 defaultDPCH-OffsetValue       |361                                 |
|-00-----  |1.1.1.1.13.2.1.2 tx-DiversityMode              |noDiversity                         |
|1.1.1.1.14 dl-InformationPerRL-List                                                            |
|1.1.1.1.14.1 dL-InformationPerRL                                                               |
|1.1.1.1.14.1.1 modeSpecificInfo                                                                |
|1.1.1.1.14.1.1.1 fdd                                                                           |
|1.1.1.1.14.1.1.1.1 primaryCPICH-Info                                                           |
|***b9***  |1.1.1.1.14.1.1.1.1.1 primaryScramblingCode     |451                                 |
|1.1.1.1.14.1.2 dl-DPCH-InfoPerRL                                                               |
|1.1.1.1.14.1.2.1 fdd                                                                           |
|0-------  |1.1.1.1.14.1.2.1.1 pCPICH-UsageForChannelEst   |mayBeUsed                           |
|***b8***  |1.1.1.1.14.1.2.1.2 dpch-FrameOffset            |122                                 |
|1.1.1.1.14.1.2.1.3 dl-ChannelisationCodeList                                                   |
|1.1.1.1.14.1.2.1.3.1 dL-ChannelisationCode                                                     |
|1.1.1.1.14.1.2.1.3.1.1 sf-AndCodeNumber                                                        |
|***b8***  |1.1.1.1.14.1.2.1.3.1.1.1 sf256                 |4                                   |
|-000----  |1.1.1.1.14.1.2.1.4 tpc-CombinationIndex        |0                                   |

+---------+------------------------------------------------+------------------------------------+
|BITMASK  |ID Name                                         |Comment or Value                    |
+---------+------------------------------------------------+------------------------------------+
|UNI SSCOP (SSCOP)   SD (= Seq. Conn.mode Data)                                                 |
|Seq. Conn.mode Data                                                                            |
|***B2***  |Padding                                        |0                                   |
|10------  |PAD length                                     |2                                   |
|--00----  |Reserved                                       |0                                   |
|----1000  |PDU Type                                       |8                                   |
|***B3***  |this Sequencenumber - N(S)                     |192                                 |
|TS 25.433 V3.9.0 (2002-03) (NBAP)   initiatingMessage (= initiatingMessage)                    |
|nbapPDU                                                                                        |
|1 initiatingMessage                                                                            |
|1.1 procedureID                                                                                |
|00011010  |1.1.1 procedureCode                            |id-radioLinkRestoration             |
|-10-----  |1.1.2 ddMode                                   |common                              |
```

16

UMTS – Signaling & Protocol Analysis

```
|---01---  |1.2 criticality                                |ignore                              |
|-----1--  |1.3 messageDiscriminator                       |dedicated                           |
|1.4 transactionID                                                                              |
|***b7***  |1.4.1 shortTransActionId                       |13                                  |
|1.5 value                                                                                      |
|1.5.1 protocolIEs                                                                              |
|1.5.1.1 sequence                                                                               |
|***B2***  |1.5.1.1.1 id                                   |id-CRNC-CommunicationContextID      |
|01------  |1.5.1.1.2 criticality                          |ignore                              |
|***B3***  |1.5.1.1.3 value                                |121817                              |
|1.5.1.2 sequence                                                                               |
|***B2***  |1.5.1.2.1 id                                   |id-Reporting-Object-RL-RestoreInd   |
|01------  |1.5.1.2.2 criticality                          |ignore                              |
|1.5.1.2.3 value                                                                                |
|1.5.1.2.3.1 rL-Set                                                                             |
|1.5.1.2.3.1.1 rL-Set-InformationList-RL-RestoreInd                                             |
|1.5.1.2.3.1.1.1 sequenceOf                                                                     |
|***B2***  |1.5.1.2.3.1.1.1.1 id                           |id-RL-Set-InformationItem-RL-Restore|
|01------  |1.5.1.2.3.1.1.1.2 criticality                  |ignore                              |
|1.5.1.2.3.1.1.1.3 value                                                                        |
|--00001-  |1.5.1.2.3.1.1.1.3.1 rL-Set-ID                  |1                                   |

+---------+------------------------------------------------+------------------------------------+
|BITMASK  |ID Name                                         |Comment or Value                    |
+---------+------------------------------------------------+------------------------------------+
|UNI SSCOP (SSCOP)   SD (= Seq. Conn.mode Data)                                                 |
|Seq. Conn.mode Data                                                                            |
|***B2***  |Padding                                        |0                                   |
|10------  |PAD length                                     |2                                   |
|--00----  |Reserved                                       |0                                   |
|----1000  |PDU Type                                       |8                                   |
|***B3***  |this Sequencenumber - N(S)                     |192                                 |
|TS 25.433 V3.9.0 (2002-03) (NBAP)   initiatingMessage (= initiatingMessage)                    |
|nbapPDU                                                                                        |
```

17

UMTS – Signaling & Protocol Analysis

```
|1 initiatingMessage                        |
|1.1 procedureID                            |
|00011010 |1.1.1 procedureCode              |id-radioLinkRestoration
|-10----- |1.1.2 ddMode                     |common
|---01--- |1.2 criticality                  |ignore
|-----1-- |1.3 messageDiscriminator         |dedicated
|1.4 transactionID                          |
|***b7*** |1.4.1 shortTransActionId         |16
|1.5 value                                  |
|1.5.1 protocolIEs                          |
|1.5.1.1 sequence                           |
|***B2*** |1.5.1.1.1 id                     |id-CRNC-CommunicationContextID
|01------ |1.5.1.1.2 criticality            |ignore
|***B3*** |1.5.1.1.3 value                  |131813
|1.5.1.2 sequence                           |
|***B2*** |1.5.1.2.1 id                     |id-Reporting-Object-RL-RestoreInd
|01------ |1.5.1.2.2 criticality            |ignore
|1.5.1.2.3 value                            |
|1.5.1.2.3.1 rL-Set                         |
|1.5.1.2.3.1.1 rL-Set-InformationList-RL-RestoreInd
|1.5.1.2.3.1.1.1 sequenceOf                 |
|***B2*** |1.5.1.2.3.1.1.1.1 id             |id-RL-Set-InformationItem-RL-Restore
|01------ |1.5.1.2.3.1.1.1.2 criticality    |ignore
|1.5.1.2.3.1.1.1.3 value                    |
|--00001- |1.5.1.2.3.1.1.1.3.1 rL-Set-ID    |1


+----------+-----------------------------------------+-----------------------------------+
|BITMASK   |ID Name                                  |Comment or Value                   |
+----------+-----------------------------------------+-----------------------------------+
|TS 25.322 V3.10.0 (2002-03) reassembled (RLC reasm.) AM DATA DCH (= Acknowledged Mode Data DCH)
|Acknowledged Mode Data DCH
|          |FP:   VPI/VCI/CID                        |"14/06/64"
|          |FP:   Direction                          |Uplink
|          |FP:   Transport Channel Type             |DCH (Dedicated Channel)
```

(18)

UMTS – Signaling & Protocol Analysis

```
|          |MAC:  Target Channel Type                |DCCH (Dedicated Control Channel)
|          |MAC:  C/T Field                          |Logical Channel 3
|          |MAC:  RLC Mode                           |Acknowledge Mode
|**B25***  |RLC:  Whole Data                         |4a 08 00 08 20 00 41 91 a0 51 06...
|TS 29.331 DCCH-UL (2002-03) (RRC_DCCH_UL)   rrcConnectionSetupComplete (= rrcConnectionSetupComplete)
|uL-DCCH-Message
|1 message
|1.1 rrcConnectionSetupComplete
|-00----- |1.1.1 rrc-TransactionIdentifier           |0
|1.1.2 startList
|1.1.2.1 sTARTSingle
|-----0-- |1.1.2.1.1 cn-DomainIdentity               |cs-domain
|**b20*** |1.1.2.1.2 start-Value                     |'000000000000001000'B
|1.1.2.2 sTARTSingle
|--1----- |1.1.2.2.1 cn-DomainIdentity               |ps-domain
|**b20*** |1.1.2.2.2 start-Value                     |'000000000000100000'B
|1.1.3 ue-RadioAccessCapability
|         |1.1.3.1 accessStratumReleaseIndicator     |r99
|1.1.3.2 pdcp-Capability
|1------- |1.1.3.2.1 losslessSRNS-RelocationSupport  |1
|1.1.3.2.2 supportForRfc2507
|         |1.1.3.2.2.1 notSupported                  |0
|1.1.3.3 rlc-Capability
|--010--- |1.1.3.3.1 totalRLC-AM-BufferSize          |kb50
|-----0-- |1.1.3.3.2 maximumRLC-WindowSize           |mws2047
|***b3*** |1.1.3.3.3 maximumAM-EntityNumber          |am6
|1.1.3.4 transportChannelCapability
|1.1.3.4.1 dl-TransChCapability
|-0100--- |1.1.3.4.1.1 maxNoBitsReceived             |b5120
|***b4*** |1.1.3.4.1.2 maxConvCodeBitsReceived       |b640
|1.1.3.4.1.3 turboDecodingSupport
|--0100-- |1.1.3.4.1.3.1 supported                   |b5120
|------01 |1.1.3.4.1.4 maxSimultaneousTransChs       |e8
|000----- |1.1.3.4.1.5 maxSimultaneousCCTrCH-Count   |1
|---0011- |1.1.3.4.1.6 maxReceivedTransportBlocks    |tb32
```

UMTS – Signaling & Protocol Analysis

```
|***b4*** |1.1.3.4.1.7 maxNumberOfTFC              |tfc128                    |
|---001-- |1.1.3.4.1.8 maxNumberOfTF               |tf64                      |
|1.1.3.4.2 ul-TransChCapability                                                |
|***b4*** |1.1.3.4.2.1 maxNoBitsTransmitted        |b3840                     |
|--0000-- |1.1.3.4.2.2 maxConvCodeBitsTransmitted  |b640                      |
|1.1.3.4.2.3 turboEncodingSupport                                              |
|***b4*** |1.1.3.4.2.3.1 supported                 |b3840                     |
|---010-- |1.1.3.4.2.4 maxSimultaneousTransChs     |e8                        |
|1.1.3.4.2.5 modeSpecificInfo                                                  |
|         |1.1.3.4.2.5.1 fdd                       |0                         |
|***b4*** |1.1.3.4.2.6 maxTransmittedBlocks        |tb8                       |
|---0011- |1.1.3.4.2.7 maxNumberOfTFC              |tfc32                     |
|***b3*** |1.1.3.4.2.8 maxNumberOfTF               |tf32                      |
|1.1.3.5 rf-Capability                                                         |
|1.1.3.5.1 fddRF-Capability                                                    |
|----11-- |1.1.3.5.1.1 ue-PowerClass               |4                         |
|------00 |1.1.3.5.1.2 txRxFrequencySeparation     |mhz190                    |
|1.1.3.6 physicalChannelCapability                                             |
|1.1.3.6.1 fddPhysChCapability                                                 |
|1.1.3.6.1.1 downlinkPhysChCapability                                          |
|--010--- |1.1.3.6.1.1.1 maxNoDPCH-PDSCH-Codes     |3                         |
|***b4*** |1.1.3.6.1.1.2 maxNoPhysChBitsReceived   |b9600                     |
|-0------ |1.1.3.6.1.1.3 supportForSF-512          |0                         |
|--0----- |1.1.3.6.1.1.4 supportOfPDSCH            |0                         |
|1.1.3.6.1.1.5 simultaneousSCCPCH-DPCH-Reception                               |
|         |1.1.3.6.1.1.5.1 notSupported            |0                         |
|1.1.3.6.1.2 uplinkPhysChCapability                                            |
|----0010 |1.1.3.6.1.2.1 maxNoDPDCH-BitsTransmitted|b2400                     |
|0------- |1.1.3.6.1.2.2 supportOfPCPCH            |0                         |
|1.1.3.7 ue-MultiModeRAT-Capability                                            |
|1.1.3.7.1 multiRAT-CapabilityList                                             |
|-0------ |1.1.3.7.1.1 supportOfGSM                |0                         |
|--0----- |1.1.3.7.1.2 supportOfMulticarrier       |0                         |
|---01--- |1.1.3.7.2 multiModeCapability           |fdd                       |
|1.1.3.8 securityCapability                                                    |
```

UMTS – Signaling & Protocol Analysis

```
|**b16*** |1.1.3.8.1 cipheringAlgorithmCap          |'0000000000000001'B      |
|**b16*** |1.1.3.8.2 integrityProtectionAlgorithmCap|'0000000000000010'B      |
|1.1.3.9 ue-positioning-Capability                                             |
|-----0-- |1.1.3.9.1 standaloneLocMethodsSupported  |0                        |
|------0- |1.1.3.9.2 ue-BasedOTDOA-Supported        |0                        |
|***b2*** |1.1.3.9.3 networkAssistedGPS-Supported   |noNetworkAssistedGPS     |
|-0------ |1.1.3.9.4 supportForUE-GPS-TimingOfCellFrames|0                    |
|--0----- |1.1.3.9.5 supportForIPDL                 |0                        |
|1.1.3.10 measurementCapability                                                |
|1.1.3.10.1 downlinkCompressedMode                                             |
|------1- |1.1.3.10.1.1 fdd-Measurements            |1                        |
|1.1.3.10.2 uplinkCompressedMode                                               |
|--1----- |1.1.3.10.2.1 fdd-Measurements            |1                        |

+---------+----------------------------------------+--------------------------+
|BITMASK  |ID Name                                 |Comment or Value          |
+---------+----------------------------------------+--------------------------+
|TS 25.322 V3.10.0 (2002-03) reassembled (RLC reasm.)  AM DATA DCH (= Acknowledged Mode Data DCH)
|Acknowledged Mode Data DCH
|         |FP:  VPI/VCI/CID                        |"14/06/67"                |
|         |FP:  Direction                          |Uplink                    |
|         |FP:  Transport Channel Type             |DCH (Dedicated Channel)   |
|         |MAC: Target Channel Type                |DCCH (Dedicated Control Channel)|
|         |MAC: C/T Field                          |Logical Channel 4         |
|         |MAC: RLC Mode                           |Acknowledge Mode          |
|**B34*** |RLC: Whole Data                         |15 80 08 00 c8 40 08 16 28 03 08...|
|TS 29.331 DCCH-UL (2002-03) (RRC_DCCH_UL)  initialDirectTransfer (= initialDirectTransfer)
|uL-DCCH-Message
|1 message
|1.1 initialDirectTransfer
|1------- |1.1.1 cn-DomainIdentity                 |ps-domain                 |
|1.1.2 intraDomainNasNodeSelector
|1.1.2.1 version
|1.1.2.1.1 release99
```

UMTS – Signaling & Protocol Analysis

```
|1.1.2.1.1.1 cn-Type                                 |
|1.1.2.1.1.1.1 gsm-Map-IDNNS                         |
|1.1.2.1.1.1.1.1 routingbasis                        |
|1.1.2.1.1.1.1.1.1 localPTMSI                        |
|**b10***  |1.1.2.1.1.1.1.1.1.1 routingparameter    |'0000001000'B                                  |
|0-------  |1.1.2.1.1.1.1.2 enteredparameter        |0                                              |
|**b208**  |1.1.3 nas-Message                       |06 42 b0 40 19 40 40 c3 78 31 05...            |
|1.1.4 v3a0NonCriticalExtensions                     |
|1.1.4.1 initialDirectTransfer-v3a0ext               |
|**b20***  |1.1.4.1.1 start-Value                   |'00000000000000100'B                           |
|TS 24.008 GPRS Mobility Manageme nt V3.11.0 (GMM-DMTAP)  ATRQ (= Attach request)                |
|Attach request                                      |
|----1000  |Protocol Discriminator                  |GPRS mobility management messages              |
|0000----  |Sub-protocol discriminator              |Skip Indicator                                 |
|00000001  |Message Type                            |1                                              |
|MS Network Capability                               |
|00000010  |IE Length                               |2                                              |
|***B2***  |MS Network Capability todoCSN1          |c5 00                                          |
|Attach type + CKSN                                  |
|-----001  |Type of attach                          |GPRS only                                      |
|----0---  |Follow on Request                       |No follow on request pending                   |
|0110----  |Key sequence                            |6                                              |
|DRX Parameter                                       |
|00001101  |Split PG cycle code (1-64, respectively)|13                                             |
|-----011  |Non-DRX timer                           |max. 4 sec                                     |
|----0---  |SPLIT on CCCH                           |Not supported by MS                            |
|0000----  |CN specific CL coefficient              |CN specific CL coeff. not specified            |
|P-TMSI (allocated)                                  |
|00000101  |IE Length                               |5                                              |
|-----100  |Type of identity                        |TMSI/P-TMSI                                    |
|----0---  |Odd/Even Indicator                      |Even no of digits                              |
|1111----  |Filler                                  |15                                             |
|***B4***  |MID P-TMSI                              |c0 10 00 1b                                    |
|Routing Area Identification                         |
|----0010  |MCC digit 1                             |2                                              |
```

```
|0111----  |MCC digit 2                             |7                                              |
|----0010  |MCC digit 3                             |2                                              |
|1111----  |MNC digit 3                             |15                                             |
|----0111  |MNC digit 1                             |7                                              |
|0110----  |MNC digit 2                             |6                                              |
|***B2***  |LAC                                     |16                                             |
|00010000  |RAC                                     |16                                             |
|MS Radio Access Capability                          |
|00000101  |IE Length                               |5                                              |
|***B5***  |Radio Access Cap.                       |43 bf 00 06 00                                 |
```

```
+---------+------------------------------------------+----------------------------------------------+
|BITMASK  |ID Name                                   |Comment or Value                              |
+---------+------------------------------------------+----------------------------------------------+
|TS 25.322 V3.10.0 (2002-03) reassembled (RLC reasm.)  AM DATA DCH (= Acknowledged Mode Data DCH) |
|Acknowledged Mode Data DCH                          |
|         |     FP:  VPI/VCI/CID                    |"14/06/64"                                    |
|         |     FP:  Direction                      |Uplink                                        |
|         |     FP:  Transport Channel Type         |DCH (Dedicated Channel)                       |
|         |     MAC: Target Channel Type            |DCCH (Dedicated Control Channel)              |
|         |     MAC: C/T Field                      |Logical Channel 4                             |
|         |     MAC: RLC Mode                       |Acknowledge Mode                              |
|**B28*** |     RLC: Whole Data                     |15 00 00 00 98 28 41 13 17 93 38...           |
|TS 29.331 DCCH-UL (2002-03) (RRC_DCCH_UL) initialDirectTransfer (= initialDirectTransfer)       |
|uL-DCCH-Message                                     |
|1 message                                           |
|1.1 initialDirectTransfer                           |
|0-------  |1.1.1 cn-DomainIdentity                 |cs-domain                                     |
|1.1.2 intraDomainNasNodeSelector                    |
|1.1.2.1 version                                     |
|1.1.2.1.1 release99                                 |
|1.1.2.1.1.1 cn-Type                                 |
|1.1.2.1.1.1.1 gsm-Map-IDNNS                         |
```

UMTS – Signaling & Protocol Analysis

```
|1.1.2.1.1.1.1.1.1 routingbasis          |                                   |
|1.1.2.1.1.1.1.1.1 localPTMSI            |                                   |
|**b10*** |1.1.2.1.1.1.1.1.1.1 routingparameter |'0000000000'B                |
|0------- |1.1.2.1.1.1.1.1.1.2 enteredparameter |0                            |
|**b160** |1.1.3 nas-Message             |44 09 9a b8 98 c4 01 10 c4 7b 09... |
|1.1.4 v3a0NonCriticalExtensions         |                                   |
|1.1.4.1 initialDirectTransfer-v3a0ext   |                                   |
|**b20*** |1.1.4.1.1 start-Value         |'00000000000000100'B               |
|TS 24.008 Mobility Management V3.11.0 (MM-DMTAP)  LUREQ (= Location updating req.) |
|Location updating req.                  |                                   |
|----0101 |Protocol Discriminator        |mobility management messages       |
|0000---- |Sub-protocol discriminator    |Skip Indicator                     |
|--001000 |Message Type                  |8                                  |
|00------ |Send Sequence Number          |Message sent from the Network      |
|------10 |LUT                           |IMSI attach                        |
|-----0-- |Spare                         |0                                  |
|----0--- |Follow-On Request             |No follow-on request pending       |
|0010---- |Key sequence                  |2                                  |
|Location Area identification            |                                   |
|----0010 |MCC digit 1                   |2                                  |
|0111---- |MCC digit 2                   |7                                  |
|----0010 |MCC digit 3                   |2                                  |
|1111---- |MNC digit 3                   |15                                 |
|----0111 |MNC digit 1                   |7                                  |
|0110---- |MNC digit 2                   |6                                  |
|***B2*** |LAC                           |16                                 |
|Mobile Station Classmark 1              |                                   |
|-----111 |RF power capability           |- unknown / undefined -            |
|----1--- |A5/1 algorithm supported      |Not Available                      |
|---0---- |Cont. Early Classmark Snd     |Cont. Ear. Cl. Sd. not impl.       |
|-10----- |Revision level                |Used by MSs supp. this version     |
|0------- |Spare                         |0                                  |
|Mobile IDentity                         |                                   |
|00000101 |IE Length                     |5                                  |
|-----100 |Type of identity              |TMSI/P-TMSI                        |
```

UMTS – Signaling & Protocol Analysis

```
|----0--- |Odd/Even Indicator            |Even no of digits                  |
|1111---- |Filler                        |15                                 |
|***B4*** |MID TMSI                      |2e 00 02 ea                        |
|Mobile Station Classmark 2              |                                   |
|00110011 |IE Name                       |Mobile Station Classmark 2         |
|00000011 |IE Length                     |3                                  |
|-----111 |RF power capability           |- unknown / undefined -            |
|----1--- |A5/1 algorithm supported      |Not Available                      |
|---0---- |Cont. Early Classmark Snd     |Cont. Ear. Cl. Sd. not impl.       |
|-10----- |Revision level                |Used by MSs supp. this version     |
|0------- |Spare                         |0                                  |
|-------0 |Frequency capability          |Doesn't support E-GSM or R-GSM band|
|------0- |VGCS notific. receipt.        |not wanted                         |
|-----0-- |VBS notific. receipt.         |not wanted                         |
|----1--- |SM capability                 |Present                            |
|--01---- |SS Screening Indicator        |Ellipis & phase 2 error handle     |
|-0------ |PS capability                 |Not present                        |
|0------- |Spare                         |0                                  |
|-------0 |A5/2 algorithm supported      |Not Available                      |
|------0- |A5/3 algorithm supported      |Not Available                      |
|-----0-- |CM Service Prompt             |Netw.init.MOCM CR - not supported  |
|----0--- |SoLSA (requ. for MS supp. GSM)|ME does not support SoLSA          |
|---0---- |UCS2 treatment                |pref. for default alphabet over UCS2|
|--0----- |LCS VA capability             |No LCS value added loc. req. not. ca|
|-0------ |spare                         |0                                  |
|1------- |Classmark 3                   |MS supports options ind.in CM3     |
```

UMTS – Signaling & Protocol Analysis

- **Solutions for the Practical Exercises**

- Part 2: Signaling & Protocol Analysis on the Uu-Interface
  QUESTION: The following strings represent MAC-PDU's which have been sent or received by the UE on different TrCH's. Please identify and determine the different fields of the respective MAC-headers:

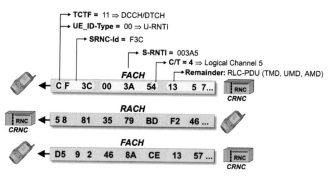

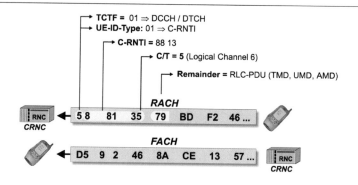

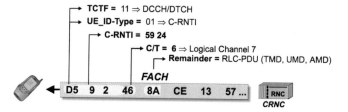

- Part 2: Signaling & Protocol Analysis on the Uu-Interface
  QUESTION: What is the configured TB-Size?
  ANSWER: The configured Transport Block Size is 21 octets (168 bit).

UMTS – Signaling & Protocol Analysis

- **Part 2: Signaling & Protocol Analysis on the Uu-Interface**

    QUESTION: Decode an RRC_CONN_REQ-message. What is the initial UE-ID type and value? What is the establishment cause?

    ANSWER: 1. The initial UE-ID-type is TMSI + LAI; ⇔ TMSI = 3E 00 1C 6B / LAI = (MCC = 2 6 2 + MNC(2 digits) = 7 6). If you count one '0'-bit too much then consider that the MNC could be 2 or 3 digits long. The respective '0'-bit indicates the first option, which is 2 digits.
    2. The establishment cause is registration.

    The decoded message looks like this:

-

```
+------+---------------------------------------+----------------------------------------+
|1.1 rrcConnectionRequest                                                               |
|1.1.1 initialUE-Identity                                                               |
|1.1.1.1 tmsi-and-LAI                                                                   |
|***B4***  |1.1.1.1.1 tmsi                     |'0011111000000000001110001101011'B      |
|1.1.1.1.2 lai                                                                          |
|1.1.1.1.2.1 plmn-Identity                                                              |
|1.1.1.1.2.1.1 mcc                                                                      |
|0010----  |1.1.1.1.2.1.1.1 digit              |2                                       |
|----0110  |1.1.1.1.2.1.1.2 digit              |6                                       |
|0010----  |1.1.1.1.2.1.1.3 digit              |2                                       |
|1.1.1.1.2.1.2 mnc                                                                      |
|***b4***  |1.1.1.1.2.1.2.1 digit              |7                                       |
|-0110---  |1.1.1.1.2.1.2.2 digit              |6                                       |
|**b16***  |1.1.1.1.2.2 lac                    |'0000000000111101'B                     |
|***b5***  |1.1.2 establishmentCause           |registration                            |
|--0-----  |1.1.3 protocolErrorIndicator       |noError                                 |
+----+--+--+--+--+--+--+--+--+--+--+--+--+--+--+--+                                     |
|HEX |0 |1 |2 |3 |4 |5 |6 |7 |8 |9 |A |B |C |D |E |F |                                  |
+----+--+--+--+--+--+--+--+--+--+--+--+--+--+--+--+                                     |
|0   |21|3e|00|1c|6b|26|23|b0|01|eb|00|00|00|00|00|                                     |
|10  |00|00|00|00|00|  |  |  |  |  |  |  |  |  |  |                                     |
+------+---------------------------------------+----------------------------------------+
```

UMTS – Signaling & Protocol Analysis

Part 5: Signaling & Protocol Analysis on the Uu-Interface

QUESTION: Determine the messages that belong together and put their numbers into the scenario underneath

ANSWER: 1. The numbers are in consecutive order: (1), (4), (5), (7), (9), (10), (11), (15), (17), (18), (20). This message sequence and the respective numbers are also indicated in the figure on the following page.

2. The most important identifiers to link the different messages to each other are:
   a) TMSI = 2E 00 02 EA $_{hex}$, the TMSI ties the RRC_CONN_SETUP to the respective RRC_CONN_REQ-message. It can also help to relate the first NAS-message to be sent (if applicable) to the respective RRC_CONN_REQ-message.
   b) DCH-Id on Iub = VPI / VCI / CID = 14/06/64. The DCH-Id can be identified through the correct ALCAP: ERQ-message. In this exercise, a DCH-TrCH is used for the registration. This DCH-Id also serves as identifier for message to/from the UE; if the registration would occur on common TrCH's, the identification of messages to/from the UE would be based on the C-RNTI / U-RNTI in the MAC-header).
   c) CRNC-Communication Context Id = 131813 = NodeB-Communication Context ID
   d) Binding-ID in NBAP: SUCC_OUT / RL_SETUP-message = 24 00 00 00 $_{hex}$ = Served User Generator Reference (603.979.776$_{dec}$) in ALCAP: ERQ-message.
   e) The ALCAP: ERQ-message is tied to the ALCAP: ECF-message through the ERQ: Originating Signaling Association Id = 33 555 271 (seen from the RNC's perspective) and the ECF: Originating Signaling Association Id = 17 (seen from the NodeB's perspective). Note: When the DCH is being released, the ALCAP: REL-message will relate to the Destination Signaling Association Id = 17 (which was originally allocated by the NodeB).
   f) The UL-Scrambling Code = 1084104 which ties together the INIT_MESS: RL_SETUP-message and the RRC_CONN_SETUP-message.

UMTS – Signaling & Protocol Analysis

Call Trace Practical Exercise / Solution:

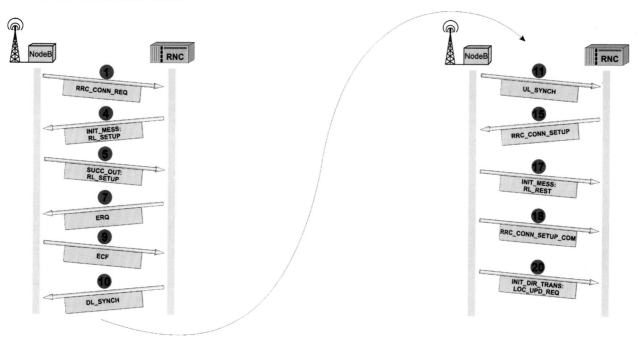

UMTS – Signaling & Protocol Analysis

▪ References

[1] GSM – Networks: Protocols, Terminology and Implementation
     ISBN 0-89006-471-7

[2] GPRS – Signaling & Protocol Analysis / Volume 1: RAN and Mobile Station
     ISBN: 1-58053-575-5

[3] EGPRS – Design Details & System Engineering
     ISBN: 1-580-536-84-0

[4] UMTS – Design Details & System Engineering
     ISBN: 3-93627-306-5

[5] GPRS – Signaling & Protocol Analysis / Volume 2: The Core Network
     ISBN: 1-58053-576-3

[6] GPRS – Network Optimization & Trouble Shooting
     ISBN 1-58053-640-9

[7] The INACON Online Encyclopaedia at www.inacon.com

[8] UMTS – Signaling & Protocol Analysis / Volume 2: The Core Network
     ISBN: ?

UMTS – Signaling & Protocol Analysis

▪ List of Acronyms

Term	Explanation
16-QAM	16 symbols Quadrature Amplitude Modulation (⇔ 3GTS 25.213)
2B1Q	Two Binary One Quaternary (⇔ Line Coding used on the ISDN U-Interface)
3G ...	3rd Generation ...
3GPP	Third Generation Partnership Project (Collaboration between different standardization organizations (e.g. ARIB, ETSI) to define advanced mobile communications standards, responsible for UMTS)
3GPP2	Third Generation Partnership Project 2 (similar to 3GPP, but consisting of ANSI, TIA and EIA-41, responsible for cdma2000, EvDO and EVDV)
8-PSK	8 Symbol Phase Shift Keying
AA	Anonymous Access
AAL-2	ATM Adaptation Layer 2 (for real-time services) (⇔ ITU-T I.363.2)
AAL-5	ATM-Adaptation Layer 5 (non-real time) (⇔ ITU-T I.363.5)
A-Bit	Acknowledgement Request Bit (⇔ used in LLC-protocol ⇔ Logical Link Control)
ABM	Asynchronous Balanced Mode
ACC	Access Control Class (⇔ 3GTS 22.011)
ACCH	Associated Control Channel (GSM / can be an SACCH or an FACCH)
ACK	Acknowledgement (⇔ 3GTS 25.214)
ACS	Active Codec Set
ADM	Asynchronous Disconnected Mode
ADPCM	Adaptive Differential Pulse Code Modulation
AES	Advanced Encryption Standard
AESA	ATM End System Address
AG	Absolute Grant (⇔ 3GTS 25.309)
AGCH	Access Grant Channel (GSM)
AH	Authentication Header (⇔ RFC 2402)
AI	Acquisition Indicator
AICH	Acquisition Indicator Channel (UMTS Physical Channel)
AK	Anonymity Key (⇔ 3GTS 33.102)
AKA	Authentication and key agreement (⇔ 3GTS 33.102)
ALCAP	Access Link Control Application Part (⇔ ITU-T Q.2630.1 / Q.2630.2)
AM	Acknowledged Mode operation (⇔ e.g. in UMTS-RLC)
AM	Amplitude Modulation
AMC	Adaptive Modulation and Coding (⇔ 3GTS 25.858)
AMD	Acknowledged Mode Data (⇔UMTS RLC PDU-type)
AMF	Authentication management field (⇔ 3GTS 33.102)
AMI	Alternate Mark Inversion (⇔ Line Coding)
AMPS	Advanced Mobile Phone System
AMR	Adaptive Multirate Encoding (⇔ 3GTS 26.090)
ANSI	American National Standards Institute
AP	Access Preamble

AP-AICH	CPCH Access Preamble Acquisition Indicator Channel (⇔ UMTS Physical Channel)
API	Access Preamble Acquisition Indicator
APN	Access Point Name (⇔ Reference to a GGSN)
APP	A Posteriori Probability (⇔ Turbo Decoding)
ARFCN	Absolute Radio Frequency Channel Number
ARIB	Association of Radio Industries and Businesses (Japanese)
ARP	Address Resolution Protocol (⇔ RFC 826)
ARQ	Automatic Repeat Request
AS	Application Server
AS	Access Stratum (⇔ UMTS)
ASC	Access Service Class
ASCI	Advanced Speech Call Items (⇔ GSM-R)
ASCII	American Standard Code for Information Interchange
ASIC	Application Specific Integrated Circuit
AS-ILCM	Application Server - Incoming Leg Control Model
ASN.1	Abstract Syntax Notation 1 (⇔ ITU-T X.680 / X.681)
AS-OLCM	Application Server - Outgoing Leg Control Model
AT-Command	Attention-Command
ATM	Asynchronous Transfer Mode (⇔ ITU-T I.361)
AuC	Authentication Center
AUTN	Authentication Token (⇔ 3GTS 33.102)
AV	Authentication Vector (⇔ 3GTS 33.102)
B8ZS	Bipolar with Eight-Zero Substitution (⇔ Line Code used at the T1-Rate (1.544 Mbit/s))
BB	Base Band module
BC	Broadcast
BCC	Base Station Color Code
BCCH	Broadcast Control Channel (UMTS Logical Cannel)
BCCH	Broadcast Control Channel (⇔ GSM Logical Channel)
BCH	Broadcast Channel (UMTS Transport Channel)
BCTP	Bearer Control Tunneling Protocol (⇔ ITU-T Q.1990)
BEC	Backward Error Correction
BEG	BEGin Message (⇔ TCAP)
BER	Bit Error Rate
BFI	Bad Frame Indication
BG	Border Gateway
BGCF	Breakout Gateway Control Function
BIB	Backward Indicator Bit
BICC	Bearer Independent Call Control (⇔ ITU-T Q.1902.1 – Q.1902.6)
BLER	Block Error Rate
BMC	Broadcast / Multicast Control (⇔ 3GTS 25.324)
BM-IWF	Broadcast Multicast Interworking Function
BQA	Bluetooth Qualification Administer
BQB	Bluetooth Qualification Body
BQRB	Bluetooth Qualification Review Board
BQTF	Bluetooth Qualification Test Facility
BS	Base Station

UMTS – Signaling & Protocol Analysis

BS_CV_MAX	Maximum Countdown Value to be used by the mobile station (⇔ Countdown Procedure)	CCITT	Comité Consultatif International Télégraphique et Téléphonique (International Telegraph and Telephone Consultative Committee)
BSC	Base Station Controller		
BSIC	Base Station Identity Code	CCM	Common Channel Management (Protocol Part on the GSM Abis-Interface / 3GTS 48.058)
BSN	Block Sequence Number (⇔ RLC) / Backward Sequence Number (⇔ SS7)	CCN	Cell Change Notification (related to Network Assisted Cell Change / 3GTS 44.060)
BSS	Base Station Subsystem		
BSSAP	Base Station Subsystem Application Part	CCPCH	Common Control Physical Channel (see also P-CCPCH and S-CCPCH)
BSSGP	Base Station System GPRS Protocol	CCS7	Common Channel Signaling System No. 7 (⇔ ITU-T Q-series of specifications, in particular Q.700 – Q.703)
BSSMAP	Base Station Subsystem Mobile Application Part (⇔ 3GTS 48.008)		
BTAB	Bluetooth Technical Advisory Board	CCTrCH	Coded Composite Transport Channel (UMTS)
BTS	Base Transceiver Station	CCTrCH	Coded Composite Transport Channel (UMTS)
BVCI	BSSGP Virtual Connection Identifier	CCU	Channel Codec Unit
C/R-Bit	Command / Response Bit	CD/CA-ICH	Collision Detection / Channel Assignment Indicator Channel (UMTS Physical Channel)
C/T-Field	logical Channel / Transport channel identification Field	CDI	Collision Detection Indicator
CAI	Channel Assignment Indicator	CDMA	Code Division Multiple Access
CAP	CAMEL Application Part (⇔ CCS7)	CDR	Call Detail Record
CBC	Cell Broadcast Center	CEPT	Conférence Européenne des Postes et Télécommunications
CBCH	Cell Broadcast Channel (GSM)	CFN	Connection Frame Number
CC	Call Control	CG	Charging Gateway
CCC	CPCH Control Command	CGF	Charging Gateway Function
CCCH	Common Control Channel (UMTS Logical Channel)	CGI	Cell Global Identification
CCCH	Common Control Channel (GSM Logical Channel)	CHAP	Challenge Handshake Authentication Protocol (⇔ RFC 1334)
CCH	Control Channel	CIC	Circuit Identity Code (⇔ ISUP)
		CIC	Call Instance Code (⇔ BICC)

UMTS – Signaling & Protocol Analysis

CID	Channel Identity (⇔ ATM)	CSPDN	Circuit Switched Public Data Network
CIDR	Classless Inter-Domain Routing (⇔ RFC 1519)	CS-X	Coding Scheme (1 – 4)
CIO	Cell Individual Offset (⇔ 3GTS 25.331)	CTCH	Common Traffic Channel (Logical) ⇔ PTM
CK	Ciphering Key	CTFC	Calculated Transport Format Combination (⇔ 3GTS 25.331)
CKSN	Ciphering Key Sequence Number		
CMC	Codec Mode Command	CV	Countdown Value
CMI	Codec Mode Indication	CW	Code Word
CMR	Codec Mode Request	cwnd	Congestion window
CN	Core Network	dBm	$X [dBm] = 10 \times \log_{10}(X [W] / 0.001 [W])$
CON	CONtinue Message (⇔ TCAP)	DBP	Diameter Base Protocol (⇔ RFC 3588)
COPS	Common Open Policy Service Protocol (⇔ RFC 2748)	DCCH	Dedicated Control Channel (UMTS Logical Channel)
		DCH	Dedicated Channel (Transport)
CPCH	Common Packet Channel (UMTS Transport Channel) ⇔ FDD only	DCM	Dedicated Channel Management (Protocol Part on the GSM Abis-Interface / 3GTS 48.058)
CPCS	Common Part Convergence Sublayer	DCS	Digital Communication System
CPICH	Common Pilot Channel (UMTS Physical Channel / see also P-CPICH and S-CPICH)	DDI	Data Description Indicator (⇔ 3GTS 25.309, 25.331)
CPS	Coding and Puncturing Scheme	DES	Data Encryption Standard
CQI	Channel Quality Indicator (⇔ 3GTS 25.214)	DHCP	Dynamic Host Configuration Protocol (⇔ RFC 2131)
CRNC	Controlling RNC	Digit	4 bit
CS	Coding Scheme	DL	Downlink
C-SAP	Control Service Access Point	DLR	Destination Local Reference (⇔ SCCP term)
CSCF	Call Session Control Function (⇔ SIP)	DNS	Domain Name System
CSD	Circuit Switched Data	DPC	Destination Point Code
CSICH	CPCH Status Indicator Channel (UMTS Physical Channel)	DPCCH	Dedicated Physical Control Channel (UMTS Physical Channel)
CSMA-CA	Carrier-Sense Multiple Access – Collision Avoidance	DPCH	Dedicated Physical Channel (UMTS / Term to combine DPDCH and DPCCH)

UMTS – Signaling & Protocol Analysis

DPDCH	Dedicated Physical Data Channel (UMTS Physical Channel)	EGPRS	Enhanced General Packet Radio Service
DRNC	Drift Radio Network Controller	E-GSM	Extended GSM (GSM 900 in the Extended Band)
DRX	Discontinuous Reception	E-HICH	E-DCH HARQ Acknowledgement Indicator Channel (⇔ 3GTS 25.211)
DS-CDMA	Direct Sequence Code Division Multiple Access	EIA	Electronic Industries Alliance (US-organization to support US industry)
DSCH	Downlink Shared Channel (UMTS Transport Channel)	EIR	Equipment Identity Register
DSL	Digital Subscriber Line	EIRENE	European Integrated Railway Radio Enhanced Network (⇔ GSM-R)
DSN	Digital Switching Network	eMLPP	enhanced Multi-Level Precedence and Pre-emption (⇔ 3GTS 23.067)
DSS1	Digital Subscriber Signaling System No.1 (⇔ also referred to as LAPD-signaling / ITU-T Q.931)	END	END Message (⇔ TCAP)
DTAP	Direct Transfer Application Part	E-RGCH	E-DCH Relative Grant Channel (⇔ 3GTS 25.211)
DTCH	Dedicated Traffic Channel (UMTS Logical Channel)	E-RNTI	E-DCH Radio Network Temporary Identifier (⇔ 3GTS 25.401)
DTM	Dual Transfer Mode (⇔ 3GTS 43.055)	ESN	Electronic Serial Number (North American Market)
DTX	Discontinuous Transmission	ESP	Encapsulating Security Payload (⇔ RFC 2406)
E-AGCH	E-DCH Absolute Grant Channel (⇔ 3GTS 25.211)	E-TFC	E-DCH Transport Format Combination (⇔ 3GTS 25.309)
Ec/No	Received energy per chip / power density in the band	Ethernet	Layer 2 Protocol for IP (⇔ IEEE 802.3)
ECSD	Enhanced Circuit Switched Data (⇔ HSCSD + EDGE)	ETSI	European Telecommunications Standard Institute
E-DCH	Enhanced Uplink Dedicated Transport Channel (⇔ 3GTS 25.211, 25.309)	EvDO	Evolution Data Only or Evolution Data Optimized (⇔ cdma2000)
EDGE	Enhanced Data Rates for Global Evolution	EVDV	Evolution Data/Voice (⇔ cdma2000)
E-DPCCH	E-DCH Dedicated Physical Control Channel(⇔3GTS 25.211)	EVM	Error Vector Magnitude
E-DPDCH	E-DCH Dedicated Physical Data Channel(⇔ 3GTS 25.211)	FACCH	Fast Associated Control Channel (GSM)
EDR	Enhanced Data Rate (⇔ more speed with Bluetooth 2.0 (⇔ 2.0 – 3.0 Mbit/s)	FACH	Forward Access Channel (UMTS Transport Channel)
EFR	Enhanced Full Rate speech codec	FBI	Feedback Information ⇔ UMTS
		FBI	Final Block Indicator

UMTS – Signaling & Protocol Analysis

FCC	Federal Communications Commission	GK	Gatekeeper
FCCH	Frequency Correction Channel (GSM)	GMM	GPRS Mobility Management
FCS	Frame Check Sequence (CRC-Check)	G-MSC	Gateway MSC
FDD	Frequency Division Duplex	GMSC-S	Gateway MSC Server
FDDI	Fiber Distributed Data Interconnect (optical Layer 2)	GMSK	Gaussian Minimum Shift Keying
FDMA	Frequency Division Multiple Access	G-PDU	T-PDU + GTP-Header
FEC	Forward Error Correction	GPRS	General Packet Radio Service
FER	Frame Error Rate	GPRS-CSI	GPRS CAMEL Subscription Information
FFH	Fast Frequency Hopping	GPRS-SSF	GPRS Service Switching Function (⇔ CAMEL)
FH-CDMA	Frequency Hopping Code Division Multiple Access	GPS	Global Positioning System
FIB	Forward Indicator Bit	GSM	Global System for Mobile Communication
FISU	Fill In Signal Unit	GSM-R	GSM for Railways
FMC	Fixed Mobile Convergence	GSN	GPRS Support Node
FN	Frame Number	GTP	GPRS Tunneling Protocol (⇔ 3GTS 29.060)
FPB	First Partial Bitmap	GTP-C	GTP Control Plane
FR	Fullrate or Frame Relay	GTP-U	GTP User Plane
FRMR	Frame Reject	GTT	Global Text Telephony (⇔ 3GTS 23.226)
FSN	Forward Sequence Number	GTTP	GPRS Transparent Transport Protocol (⇔ 3GTS 44.018)
FTP	File Transfer Protocol (⇔RFC 959)	HARQ	Hybrid ARQ (⇔ 3GTS 25.212)
GCC	Generic Call Control	HCS	Hierarchical Cell Structure
GCF	General Certification Forum	HDB3	High Density Bipolar Three (⇔ Line Coding used for E1 (PCM 30))
GEA	GPRS Encryption Algorithm	HDLC	High level Data Link Control
GERAN	GSM EDGE Radio Access Network	HLR	Home Location Register
GGSN	Gateway GPRS Support Node		
GIF	Graphics Interchange Format		

UMTS – Signaling & Protocol Analysis

HMAC	Keyed Hashing for Message Authentication (⇔ RFC 2104)	ICM	Initial Codec Mode
H-PLMN	Home PLMN	ICMP	Internet Control Message Protocol (⇔ RFC 792)
HR	Halfrate	ICS	Implementation Conformance Statement
H-RNTI	HS-DSCH Radio Network Transaction Identifier (⇔ 3GTS 25.331, 25.433)	I-CSCF	Interrogating Call Session Control Function (⇔ SIP)
HSCSD	High Speed Circuit Switched Data	IE	Information Element
HSDPA	High Speed Downlink Packet Access (⇔ 3GTS 25.301, 25.308, 25.401, 3GTR 25.848)	IEEE	Institute of Electrical and Electronics Engineers
HS-DPCCH	High Speed Dedicated Physical Control Channel (⇔ 3GTS 25.211)	IETF	Internet Engineering Task Force (www.ietf.org)
HS-DSCH	High Speed Downlink Shared Transport Channel (⇔ 3GTS 25.211, 25.212, 25.308)	IHOSS	Internet Hosted Octet Stream Service
		IK	Integrity Key
HS-PDSCH	High Speed Physical Downlink Shared Channel (⇔ 3GTS 25.211)	IKE	Internet Key Exchange (⇔ RFC 2409)
		IKMP	Internet Key Management Protocol
HSS	Home Subscriber Server (⇔ 3GTS 23.002). HSS replaces the HLR with 3GPP Rel. 5	ILCM	Incoming Leg Control Model
		IMEI	International Mobile Equipment Identity
HS-SCCH	High Speed Shared Control Channel (⇔ 3GTS 25.211, 25.214)	IMPI	IP Multimedia Private Identity
HSUPA	High Speed Uplink Packet Access (⇔ 3GTS 25.301, 25.309, 25.401, 3GTR 25.896)	IMPU	IP Multimedia Public Identity
		IMS	Internet Protocol Multimedia Core Network Subsystem (⇔ Rel. 5 onwards)
HTTP	HyperText Transfer Protocol (⇔ RFC 2616)	IMSI	International Mobile Subscriber Identity
HUMAN	High-speed Unlicensed Metropolitan Area Network	IMT-2000	International Mobile Telecommunications for the year 2000
I+S	Information + Supervisory		
IAM	Initial Address Message (ISUP ⇔ ISDN User Part)	INAP	Intelligent Network Application Part (⇔ CCS7)
IANA	Internet Assigned Numbers Authority	IOV-I / IOV-UI	Input Offset Variable for I+S and UI-Frames (⇔ for ciphering in GPRS)
ICANN	Internet Corporation for Assigned Names and Numbers	IP	Internet Protocol (⇔ RFC 791)
ICH	Indicator Channel (UMTS Physical Channel / see also PICH, AICH, CD/CA-ICH)	IPBCP	IP Bearer Control Protocol (⇔ ITU-T Q.1970)
ICH	Indicator Channel	IPCP	Internet Protocol Control Protocol (⇔ RFC 1332)

UMTS – Signaling & Protocol Analysis

IPsec	Internet Protocol / secure (⇔ RFC 2401)	LA	Location Area
IPv4	Internet Protocol (version 4)	LAC	Location Area Code
IPv6	Internet Protocol (version 6)	LAI	Location Area Identification (LAI = MCC + MNC + LAC)
IR	Incremental Redundancy (⇔ ARQ II)	LAPB	Link Access Procedure Balanced
ISAKMP	Internet Security Association and Key Management Protocol (⇔ RFC 2408)	LAPD	Link Access Protocol for the ISDN D-Channel
ISC	IP multimedia Subsystem Service Control-Interface	LBS	Location Based Service
ISCP	Interference Signal Code Power (⇔ 3GTS 25.215 / 3GTS 25.102)	LCP	Link Control Protocol (⇔ PPP)
ISDN	Integrated Services Digital Network	LCS	LoCation Service
I-SIM	IMS capable SIM	LI	Length Indicator
ISO	International Standardization Organization	LLC	Logical Link Control-Protocol
ISP	Internet Service Provider	LPD	Link Protocol Discriminator
ISPC	International Signaling Point Code (⇔ ITU-T Q.708)	LSB	Least Significant Bit
ISUP	ISDN User Part (⇔ ITU-T Q.761 – Q.765)	LSSU	Link Status Signal Unit
ITU-T	International Telecommunication Union – Telecommunication Sector	M3UA	MTP-3 User Adaptation Layer (⇔ RFC 3332 / 3GPP 29.202 (Annex A))
Iub-FP	Iub-Frame Protocol (⇔ 3GTS 25.427 / 25.435)	MAC	Medium Access Control (UMTS ⇔ 3GTS 25.321)
Iu-FP	Iu-Frame Protocol (⇔ 3GTS 25.415)	MAC	Medium Access Control ((E)GPRS ⇔ 3GTS 04.60 / 3GTS 44.060)
Iur-FP	Iur-Frame Protocol (⇔ 3GTS 25.424, 3GTS 25.425, 25.426, 25.435)	MAC	Message Authentication Code (⇔ 3GTS 33.102)
JPEG	Joint Picture Expert Group	MAC-e	MAC-E-DCH (⇔ 3GTS 25.321)
kbps	kilo-bits per second	MAC-es	MAC-E-DCH SRNC (⇔ 3GTS 25.321)
L1	Layer 1 (physical layer)	MAC-hs	MAC-High Speed (⇔ 3GTS 25.321)
L2	Layer 2 (data link layer)	MAN	Metropolitan Area Network
L2TP	Layer 2 Tunneling Protocol (⇔ RFC 2661)	MAP	Mobile Application Part
L3	Layer 3 (network layer)	MASF	Minimum Available Spreading Factor

UMTS – Signaling & Protocol Analysis

Max [X, Y]	The value shall be the maximum of X or Y, whichever is bigger	MMCC	Multimedia Call Control
MBZ	Must Be Zero	MMS	Multimedia Messaging Service (⇔ 3GTS 22.140, 3GTS 23.140]
MCC	Mobile Country Code	MNC	Mobile Network Code
Mcps	Mega Chip Per Second	MNRG	Mobile Not Reachable for GPRS flag
MCS-X	Modulation and Coding Scheme (1 – 9) and for HSDPA / HSUPA	MOC	Mobile Originating Call
MCU	Multipoint Control Unit (⇔ H.323 equipment)	MPCC	Multiparty Call Control
MD-X	Message Digest Algorithm (MD-2, 4, 5 are defined) (MD-5 ⇔ RFC 1321)	MPEG	Motion Picture Expert Group
		MRFC	Multimedia Resource Function Controller
ME	Mobile Equipment (ME + SIM = MS)	MRFP	Multimedia Resource Function Processor
MEGACO	Media Gateway Control Protocol (⇔ ITU-T H.248 incl. Annex F – H and IETF RFC 3015)	MRU	Maximum Receive Unit (⇔ PPP)
		MRW	Move Receiving Window
MExE	Mobile Station Application Execution Environment	MS	Mobile Station
MGC	Media Gateway Controller	MSB	Most Significant Bit
MGCF	Media Gateway Control Function	MSC	Mobile Services Switching Center
MGCP	Media Gateway Control Protocol (⇔ RFC 2705)	MSC-S	MSC-Server
MGW	Media Gateway	MS-ISDN	Mobile Subscriber – International Service Directory Number
MIDI	Musical Instrument Digital Interface		
MIME	Multipurpose Internet Mail Extensions	MSS	Maximum Segment Size (⇔ TCP)
MIMO	Multiple In, Multiple Out (⇔ 3GTR 25.848)	MSU	Message Signal Unit
MIN	Mobile Identity Number (North American Market)	MT	Mobile Terminal or Mobile Terminating
Min [X, Y]	The value shall be the minimum of X or Y, which ever is smaller	MTC	Mobile Terminating Call
		MTP	Message Transfer Part (⇔ ITU-T Q.701 – Q.709)
MLP	MAC Logical Channel Priority	MTP-3b	Message Transfer Part level 3 / broadband (⇔ ITU-T Q.2210)
MLPP	Multi-Level Precedence and Pre-emption (⇔ ITU-T Q.85 / Clause 3)		
MM	Mobility Management	MTU	Maximum Transmit Unit (⇔ IP)

UMTS – Signaling & Protocol Analysis

NACC	Network Assisted Cell Change (⇔ 3GTS 44.060)	NT	Network Termination
NACK	Negative Acknowledgement (⇔ 3GTS 25.308, 25.309))	O&M	Operation and Maintenance
		Octet	8 bit
NAS	Non-Access-Stratum (⇔ UMTS)	OLCM	Outgoing Leg Control Model
NAT	Network Address Translation (⇔ RFC 1631)	OMA	Open Mobile Alliance (⇔ http://www.openmobilealliance.org/)
NBAP	NodeB Application Part (⇔ 3GTS 25.433)		
NBNS	NetBios Name Service	OMC	Operation and Maintenance Center
NC	Neighbor Cell	OoBTC	Out of Band Transcoder Control (⇔ 3GTS 23.153)
NCC	Network Color Code	OPC	Originating Point Code
NCP	Network Control Protocol (⇔ PPP)	OPWA	One Pass With Advertising (⇔ Term in RSVP)
NGN	Next Generation Networks	OSA	Open Service Access
NI	Network Indicator	OSA-SCS	Open Service Access – Service Capability Server
NIC	Network Interface Card	OSI	Open System Interconnection
NPB	Next Partial Bitmap	OSP	Octet Stream Protocol
N-PDU	Network-Protocol Data Unit (⇔ IP-Packet, X.25-Frame)	OTDOA	Observed Time Difference Of Arrival
		OVSF	Orthogonal Variable Spreading Factor
NS	Network Service	P/F-Bit	Polling/Final - Bit
NSAPI	Network Service Access Point Identifier	PABX	Private Automatic Branch Exchange
N-SAW	N-Channel Stop and Wait (⇔ 3GTS 25.309, 3GTR 25.848)	PACCH	Packet Associated Control Channel ((E)GPRS)
NSE	Network Service Entity	PAD	Packet Assembly Disassembly
NSPC	National Signaling Point Code	PAGCH	Packet Access Grant Channel ((E)GPRS)
NSS	Network Switching Subsystem	PAP	Password Authentication Protocol (⇔ RFC 1334)
NS-VC	Network Service – Virtual Connection	PBCCH	Packet Broadcast Control Channel ((E)GPRS)
NS-VCG	Network Service – Virtual Connection Group	PCCCH	Packet Common Control Channel ((E)GPRS)
NS-VL	Network Service – Virtual Link	PCCH	Paging Control Channel (UMTS Logical Channel)

UMTS – Signaling & Protocol Analysis

P-CCPCH	Primary Common Control Physical Channel (UMTS / used as bearer for the BCH TrCH)
PCH	Paging Channel (UMTS / Transport Channel)
PCH	Paging Channel (GSM / Logical Channel)
PCM	Pulse Code Modulation
PCN	Personal Communication Network
PCPCH	Physical Common Packet Channel (UMTS Physical Channel)
P-CPICH	Primary Common Pilot Channel (UMTS Physical Channel)
PCS	Personal Communication System
P-CSCF	Proxy Call Session Control Function (⇔ SIP)
PCU	Packet Control Unit
PD	Protocol Discriminator
PDCH	Packet Data Channel ((E)GPRS)
PDCP	Packet Data Convergence Protocol (⇔ 3GTS 25.323)
PDF	Policy Decision Function (⇔ Part of the IP Multimedia Subsystem)
PDH	Plesiochronous Digital Hierarchy
PDN	Packet Data Network
PDP	Packet Data Protocol
PDSCH	Physical Downlink Shared Channel (UMTS Physical Channel)
PDTCH	Packet Data Traffic Channel ((E)GPRS)
PDU	Protocol Data Unit or Packet Data Unit
PER	Packed Encoding Rules (⇔ ITU-T X.691)
PFC	Packet Flow Context
PFI	Packet Flow Identifier
PHY	Physical Layer
PICH	Page Indicator Channel (UMTS Physical Channel)
PLC	Power Line Communications
PLMN	Public Land Mobile Network
PMM	Packet Mobility Management
PN	Pseudo Noise
PNCH	Packet Notification Channel ((E)GPRS)
PoC	Push to talk over Cellular (⇔ 3GTR 29.979 and various OMA-specifications)
POP	Post Office Protocol (⇔ RFC 1939)
POTS	Plain Old Telephone Service
PPCH	Packet Paging Channel ((E)GPRS)
PPP	Point-to-Point Protocol (⇔ RFC 1661)
PRA	PCPCH Resource Availability
PRACH	Physical Random Access Channel ⇔ UMTS
PRACH	Packet Random Access Channel ((E)GPRS)
PRD	Bluetooth Qualification Program Reference Document
PRI	Primary rate access ISDN-user interface for PABX's (23 or 30 B-channels plus one D-Channel)
PS	Puncturing Scheme
PSC	Primary Synchronization Code or Primary Scrambling Code (both used in UMTS)
P-SCH	Primary Synchronization Channel (physical)
PSD	Power Spectral Density (⇔ 3GTS 25.215 / 3GTS 25.102)

UMTS – Signaling & Protocol Analysis

PSK	Phase Shift Keying
PSPDN	Packet Switched Public Data Network
PSTN	Public Switched Telephone Network
PT	Protocol Type (⇔ GTP or GTP')
PTCCH	Packet Timing Advance Control Channel ((E)GPRS)
PTCCH/D	Packet Timing Advance Control Channel / Downlink Direction ((E)GPRS)
PTCCH/U	Packet Timing Advance Control Channel / Uplink Direction ((E)GPRS)
PTM	Point to Multipoint
P-TMSI	Packet TMSI
PTP	Point to Point
PVC	Permanent Virtual Circuit
QE	Quality Estimate
QoS	Quality of Service
QPSK	Quadrature Phase Shift Keying (⇔ 3GTS 25.213)
RA	Routing Area
RAB	Radio Access Bearer
RAC	Routing Area Code
RACH	Random Access Channel (UMTS Transport Channel)
RACH	Random Access Channel (GSM)
RADIUS	Remote Authentication Dial In User Service (⇔ RFC 2865)
RAI	Routing Area Identification
RANAP	Radio Access Network Application Part (⇔ 3GTS 25.413)
RAND	Random Number
RAT	Radio Access Technology (e.g. GERAN, UTRAN, ...)
RATSCCH	Robust AMR Traffic Synchronized Control CHannel
RB	Receive Block Bitmap (⇔ EGPRS)
RB	Radio Bearer
RBB	Receive Block Bitmap (⇔ GPRS)
REJ	Reject
RF	Radio Frequency
RFC	Request for Comments (⇔ Internet Standards)
RFID	Radio Frequency Identification
RG	Relative Grant (⇔ 3GTS 25.309)
R-GSM	Railways-GSM
RL	Radio Link
RLC	Radio Link Control (UMTS ⇔ 3GTS 25.322)
RLC	Radio Link Control ((E)GPRS / 3GTS 04.60 / 3GTS 44.060)
RLM	Radio Link Management (Protocol Part on the GSM Abis-Interface / 3GTS 48.058)
RLP	Radio Link Protocol (⇔ 3GTS 24.022)
RLS	Radio Link Set (⇔ 3GTS 25.309, 25.433)
RNC	Radio Network Controller
RNL	Radio Network Layer
RNR	Receive Not Ready
RNS	Radio Network Subsystem
RNSAP	Radio Network Subsystem Application Part (⇔ 3GTS 25.423)
RNTI	Radio Network Temporary Identifier

UMTS – Signaling & Protocol Analysis

RPLMN	Registered PLMN	SACCH	Slow Associated Control Channel (GSM)
RPR	Resilient Packet Ring (⇔ IEEE 802.17)	SACCH/MD	SACCH Multislot Downlink (related control channel of TCH/FD / GSM)
RR	Radio Resource Management	SAI	Service Area Identifier
RR	Receive Ready (LAPD/LLC/RLP-Frame Type)	SAIC	Single Antenna Interference Cancellation
RRBP	Relative Reserved Block Period	SANC	Signaling Area Network Code (⇔ ITU-T Q.708)
RRC	Radio Resource Control (⇔ 3GTS 25.331)	SAP	Service Access Point
RRC-Filter	Root Raised Cosine Filter	SAPI	Service Access Point Identifier
RSC	Recursive Systematic Convolutional Coder(⇔ Turbo Coding, 25.212)	SAR	Segmentation And Reassembly (ATM-sublayer)
RSCP	Received Signal Code Power (⇔ 3GTS 25.215)	SC	Serving Cell
RSN	Retransmission Sequence Number (⇔ 3GTS 25.309)	SCCP	Signaling Connection Control Part (⇔ ITU-T Q.711 – Q.714)
RSSI	Received Signal Strength Indicator	S-CCPCH	Secondary Common Control Physical Channel (used as bearer for the FACH and PCH TrCH's / UMTS Physical Channel)
RSVP	Resource Reservation Protocol (⇔ RFC 2205)	SCH	Synchronization Channel (UMTS Physical Channel / see also P-SCH and S-SCH)
RTO	Retransmission Time Out		
RTP	Real-time Transport Protocol (⇔ RFC 3550)	SCH	Synchronization Channel (GSM)
RTT	RoundTrip Time (⇔ RFC 793)	S-CPICH	Secondary Common Pilot Channel (UMTS Physical Channel)
RV	Redundancy and Constellation Version (⇔ 3GTS 25.212)	SCR	Source Controlled Rate
RX	Receive	S-CSCF	Serving Call Session Control Function (⇔ SIP)
SA	Service Area	SCTP	Stream Control Transmission Protocol (⇔ RFC 2960)
SAAL-NNI	Signaling ATM Adaptation Layer – Network Node Interface	SDCCH	Stand Alone Dedicated Control Channel
SAB	Service Area Broadcast	SDH	Synchronous Digital Hierarchy
SABM(E)	Set Asynchronous Balanced Mode (Extended for Modulo 128 operation) (LAPD/LLC/RLP-Frame Type)	SDMA	Space Division Multiple Access
SABP	Service Area Broadcast Protocol (⇔ 3GTS 25.419)	SDU	Service Data Unit (⇔ the payload of a PDU)

UMTS – Signaling & Protocol Analysis

SF	Spreading Factor	SM	Session Management (⇔ 3GTS 23.060, 3GTS 24.008)
SFH	Slow Frequency Hopping	SMS	Short Message Service (⇔ 3GTS 24.011, 3GTS 23.040)
SFN	System Frame Number		
SG	Security Gateway (IPsec / ⇔ RFC 2401)	SM-SC	Short Message Service Center
SGSN	Serving GPRS Support Node	SMSCB	Short Message Services Cell Broadcast
SGW	Signaling Gateway (SS7 ⇔ IP)	SMS-G-MSC	SMS Gateway MSC (for Short Messages destined to Mobile Station)
SHA	Secure Hash Algorithm	SMS-IW-MSC	SMS Interworking MSC (for Short Messages coming from Mobile Station)
SHCCH	Shared Channel Control Channel (UMTS Logical Channel / ⇔ TDD only)	SMTP	Simple Mail Transfer Protocol (⇔ RFC 2821)
SI	Service Indicator	SN	Sequence Number
SIB	System Information Block	SND	Sequence Number Downlink (⇔ GTP)
SID	Silence Insertion Descriptor	SNDCP	Subnetwork Dependent Convergence Protocol
SID	Size InDex (⇔ 3GPP 25.321)	SNM	Signaling Network Management Protocol (⇔ ITU-T Q.704 (3))
SIF	Signaling Information Field		
SIG	Special Interest Group (⇔ e.g. Bluetooth)	SNN	SNDCP N-PDU Number Flag
SIM	Subscriber Identity Module	SN-PDU	Segmented N-PDU (SN-PDU is the payload of SNDCP)
SIO	Service Information Octet	SNR	Signal to Noise Ratio
SIP	Session Initiation Protocol (⇔ RFC 3261)	SNTM	Signaling Network Test & Maintenance (⇔ ITU-T Q.707)
SIR	Signal to Interference Ratio	SNU	Sequence Number Uplink (⇔ GTP)
SLC	Signaling Link Code	SOAP	Simple Object Access Protocol (⇔ http://www.w3.org/TR/2000/NOTE-SOAP-20000508)
SLF	Subscriber Locator Function		
SLR	Source Local Reference	SPC	Signaling Point Code
SLS	Signaling Link Selection	SPI	Security Parameter Index (⇔ RFC 2401)
SLTA	Signaling Link Test Acknowledge	SQN	Sequence number (used in UMTS-security architecture / 3GTS 33.102)
SLTM	Signaling Link Test Message	SRB	Signaling Radio Bearer

UMTS – Signaling & Protocol Analysis

SRES	Signed Response	SUERM	Signal Unit Error Rate Monitor (⇔ ITU-T Q.703 (10))
SRNC	Serving Radio Network Controller	SUFI	Super Field (RLC-Protocol)
SRNS	Serving Radio Network Subsystem	SVC	Switched Virtual Circuit
SRTT	Smoothed RoundTrip Time (⇔ RFC 793)	SWAP	Shared Wireless Access Protocol (⇔ Home RF)
SSC	Secondary Synchronization Code	TA	Terminal Adapter (⇔ ISDN)
SSCF	Service Specific Co-ordination Function	TA	Timing Advance
SSCF/NNI	Service Specific Coordination Function – Network Node Interface Protocol (⇔ ITU-T Q.2140)	TACS	Total Access Communication System
SSCF/UNI	Service Specific Coordination Function – User Network Interface Protocol (⇔ ITU-T Q.2130)	TAF	Terminal Adopter Function (⇔ 3GTS 27.001)
S-SCH	Secondary Synchronization Channel (physical)	TAI	Timing Advance Index
SSCOP	Service Specific Connection Oriented Protocol (⇔ ITU-T Q.2110)	TB	Transport Block
		TBF	Temporary Block Flow
SSCOPMCE	Service Specific Connection Oriented Protocol in a Multi-link or Connectionless Environment (⇔ ITU-T Q.2111)	TBS	Transport Block Set
		TCAP	Transaction Capabilities Application Part (⇔ Q.771 – Q.773)
SSCS	Service Specific Convergence Sublayer	TCH	Traffic Channel
SSDT	Site Selection Diversity Transmission	TCH/FD	Traffic Channel / Fullrate Downlink
SSN	Start Sequence Number (⇔ related to ARQ-Bitmap in GPRS / EGPRS)	TCH-AFS	Traffic CHannel Adaptive Full rate Speech
SSN	Send Sequence Number (⇔ GSM MM and CC-Protocols)	TCH-AHS	Traffic Channel Adaptive Half rate Speech
		TCP	Transmission Control Protocol
SSSAR	Service Specific Segmentation And Reassembly (⇔ ITU-T I.366.1)	TCTF	Target Channel Type Field
ssthresh	Slow start threshold (⇔ RFC 2001)	TCTV	Transport Channel Traffic Volume
STC	Signaling Transport Converter on MTP-3 and MTP-3b (⇔ ITU-T Q.2150.1) / Signaling Transport Converter on SSCOP and SSCOPMCE (⇔ ITU-T Q.2150.2)	TDD	Time Division Duplex
		TDMA	Time Division Multiple Access
		TE	Terminal Equipment
STTD	Space Time block coding based Transmission Diversity	TEID	Tunnel Endpoint Identifier (⇔ GTP / 3GTS 29.060)

UMTS – Signaling & Protocol Analysis

TF	Transport Format	TM	Transparent Mode operation (⇔ UMTS-RLC)
TFC	Transport Format Combination	TM	Transmission Modules
TFCI	Transport Format Combination Identifier	TMD	Transparent Mode Data (⇔UMTS RLC PDU-type)
TFCS	Transport Format Combination Set	TMSI	Temporary Mobile Subscriber Identity
TFI	Transport Format Indication (⇔ UMTS)	TNL	Transport Network Layer (⇔ 3GTS 25.401)
TFI	Temporary Flow Identity (⇔ (E)GPRS)	TPC	Transmit Power Command
TFO	Tandem Free Operation (⇔ 3GTS 22.053)	T-PDU	Payload of a G-PDU which can be user data, i.e. possibly segmented IP-frames, or GTP signaling information (⇔ GTP)
TFRC	Transport Format and Resource Combination (⇔ 3GTS 25.308)		
TFRI	Transport Format and Resource Indicator (<=> 3GTS 25.308, 25.321)	TQI	Temporary Queuing Identifier
		TRAU	Transcoder and Rate Adaption Unit
TFS	Transport Format Set	TrCH	Transport Channel (UMTS)
TGD	Transmission Gap start Distance (⇔ 3GTS 25.215)	TrFO	Transcoder Free Operation
TGL	Transmission Gap Length (⇔ 3GTS 25.215)	TrGW	Transition Gateway (IPv4 ⇔ IPv6)
TGPRC	Transmission Gap Pattern Repetition Count (⇔ 3GTS 25.215)	TRX	Transmitter / Receiver
TGSN	Transmission Gap Starting Slot Number (⇔ 3GTS 25.215)	TS	Timeslot
		TSC	Training Sequence Code
TH-CDMA	Time Hopping Code Division Multiple Access	TSN	Transmission Sequence Number (⇔ 3GTS 25.321)
THIG	Topology Hiding Inter Network Gateway	TSTD	Time Switched Transmit Diversity
TI	Transaction Identifier	TTI	Transmission Time Interval
TIA	Telecommunications Industry Association	TTL	Time To Live (⇔ IP-Header / RFC 791)
TID	Tunnel Identifier	TX	Transmit
TLLI	Temporary Logical Link Identifier	UA	User Agent
TLS	Transport Layer Security (⇔ RFC 2246 / RFC 3546 / formerly known as SSL or Secure Socket Layer)	UA	Unnumbered Acknowledgement (LAPD/LLC/RLP-Frame Type)
TLV	Tag / Length / Value Notation	UAC	User Agent Client

UMTS – Signaling & Protocol Analysis

UARFCN	UMTS Absolute Radio Frequency Channel Number	UTRA	UMTS Terrestrial Radio Access
UART	Universal Asynchronous Receiver and Transmitter	UTRAN	UMTS Terrestrial Radio Access Network
UAS	User Agent Server	UUI	User to User Information
UDP	User Datagram Protocol (⇔ RFC 768)	UUS	User-User-Signaling (⇔ 3GTS 23.087)
UE	User Equipment	UWB	Ultra-Wide Band
UEA	UMTS Encryption Algorithm (⇔ 3GTS 33.102)	UWC	Universal Wireless Convergence (Merge IS-136 with GSM)
UI	Unnumbered Information (⇔ LAPD) / Unconfirmed Information (⇔ LLC) / Frame Type	VAD	Voice Activity Detector
UIA	UMTS Integrity Algorithm (⇔ 3GTS 33.102)	VBS	Voice Broadcast Service (⇔ GSM-R)
UICC	Universal Integrated Circuit Card (⇔ 3GTS 22.101 / Bearer card of SIM / USIM)	VC	Virtual Circuit
		VCI	Virtual Circuit Identifier (⇔ ATM)
UL	Uplink	VGCS	Voice Group Call Service (⇔ GSM-R)
UM	Unacknowledged Mode operation (⇔ UMTS-RLC)	VHE	Virtual Home Environment (⇔ 3GTS 22.121, 3GTS 23.127)
UMD	Unacknowledged Mode Data (⇔UMTS RLC PDU-type)	VLR	Visitor Location Register
UMTS	Universal Mobile Telecommunication System	VPI	Virtual Path Identifier (⇔ ATM)
URA	UTRAN Registration Area	V-PLMN	Visited PLMN
URI	Uniform Resource Identifier	VPN	Virtual Private Network
URL	Uniform Resource Locator (⇔ RFC 1738)	WAP	Wireless Application Protocol
USAT	USIM Application Toolkit	WCDMA	Wide-band Code Division Multiple Access
USB	Universal Serial Bus	WIMAX	Worldwide Interoperability for Microwave Access (⇔ IEEE 802.16)
USCH	Uplink Shared Channel (UMTS Transport Channel ⇔ TDD only)	WINS	Windows Internet Name Service
USF	Uplink State Flag	W-LAN	Wireless Local Area Network (⇔ IEEE 802.11)
USIM	Universal Subscriber Identity Module (⇔ 3GTS 31.102)	WMAN	Wireless Metropolitan Area Network
UTF-8	Unicode Transformation Format-X (Is an X-bit) lossless encoding of Unicode characters)	WSN	Window Size Number

UMTS – Signaling & Protocol Analysis

XID	Exchange Identification (LAPD/LLC-Frame Type)	XOR	Exclusive-Or Logical Combination

Index:

A

AAL2L3 24
Access Link Control Application Part 24
access service class 106
Access Stratum 12, 16
Acknowledged Mode 118
ACK-SUFI 140
Acquisition Indicator Channel 64
Adaptive Multirate Encoding 14
AICH 64
AICH Transmission Timing 116
ALCAP 24
AM 118
AMD-PDU 136
AMR 14
AP-AICH 64
AS 12, 16
ASC 106
ASN.1 .... 172, 180
ASN.1 Types 172

B

Basic Encoding Rules 172
BCCH 36, 72
BCH 80
BER 172, 224

C

BG 6
Binding ID 396
bit mode 122
BITMAP-SUFI 144
BMC 20
BO 100
Border Gateway 6
broadcast 20
Broadcast Control Channel 36
Broadcast Domain 10, 20
buffer occupancy 100

C/T-field 84
CBC 6, 20
CCCH 36, 76
CCTrCH 38, 48
CD/CA-ICH 64
Cell Broadcast Center 6
cell update 358
CELL_DCH-State 162
CELL_FACH-State 164
CELL_PCH-State 166
CG 6
channel mapping 72
channelization code 28
channels 32
chips 28

Circuit-Switched Control Plane 12
Circuit-Switched Core Network 6
Circuit-Switched User Plane 14
Class Field 174
Coded Composite Transport Channel 48
Common Control Channel 36
Common Traffic Channel 36
common transport channels 38
control channels 36
Controlling RNC 8
CPCH Access Preamble Acquisition Indicator Channel 64
CPCH Collision Detection / Channel Assignment Indicator Channel 64
CPCH Status Indicator Channel 64
CPICH Ec/No 222, 312, 320, 326, 348
CPICH RSCP 222, 312, 320, 326
CRC-check 120
CRNC 8
CRNC-Communication Context ID 396
C-RNTI 86, 96, 358, 360, 364
CSICH 64
CTCH 36, 78

D

D/C-bit 136
DCCH 36, 74, 84
DCH 38, 84
Dedicated Channel 38
Dedicated Control Channel 36
Dedicated Physical Control Channel 58
Dedicated Physical Data Channel 58
Dedicated Traffic Channel 36
dedicated transport channels 38
Destination Signaling Association ID 396

D-field 52
DPCCH/D 60
DPCCH/D for CPCH 60
DPCCH/U 58
DPCH 60
DPDCH/D 60
DPDCH/U 58
Drift RNC 8
Drift-RNC 22
DRNC 8
D-RNTI 86, 96, 360, 364
DSCH 84
DSCH-RNTI 86
DTCH 36, 74, 84
DTX 28, 30
dynamic part 40
dynamic persistence value 106

E

E-bit 124
E-interface 11
EIR 6
Equipment Identity Register 6
event ID's 216

F

FBI-field 52
Format Field 174

G

GERAN 6

UMTS – Signaling & Protocol Analysis

G-MSC 6
Gn-interface 15, 17

H

handover 162
handover reference 338
hard handover 196
HE-field 136
HFNI 148
HND_UTRAN_CMD 350
Hyper Frame Number Indicator 148

I

I/Q-Code Multiplexing 30
Initial UE-Identity 396
Iu-bc-interface 20
Iub-interface 6, 11, 15, 17, 20
Iu-cs-interface 6, 11
Iu-ps-interface 6, 15, 16, 17
Iur-interface 8, 22

L

legend 230
Length Field 174
length indicator 124, 136
LIST-SUFI 140
logical channels 36

M

M3UA 16

MAC logical channel priority 106
MAC-b 94
MAC-c/sh 96
MAC-d 98
MAC-header 80
MAC-PDU 80
macrodiversity 98
mapping of logical channels 72
MaxDAT (RLC) 148, 150
MaxWRW 148
MEAS_INFO 348
measurements 216
message descriptors 230
MLP 106
modulation 28
MRW-ACK-SUFI 144
MRW-SUFI 144
MSC 6
MTP-3 User Adaptation 16
MTP-3b 16

N

N308 .... 206, 256, 280, 304
NAS 12, 16
neighbor cell 94
NO_MORE-SUFI 140
NodeB 6
NodeB-Communication Context ID 396
Non Access Stratum 12, 16

O

octet mode 122

UMTS – Signaling & Protocol Analysis

Originating Signaling Association ID 396

P

Packed Encoding Rules 180
Packet-Switched Control Plane 16
Packet-Switched Core Network 6
Packet-Switched User Plane 18
PAG_TYPE1 .... 166, 170, 358
PAG_TYPE2 .... 162, 164, 170
paging / paging response in CELL_FACH-, CELL_DCH-state 288
paging / paging response in CELL_PCH-, URA_PCH-state 288
paging / paging response in RRC-idle state 284
Paging Control Channel 36
Paging Indicator Channel 62
P-bit 136
PCCH 36, 72
P-CCPCH 62
PCPCH 58
P-CPICH 62
PDSCH 60
PER 180
periodical cell update 358
persistence 106
persistence value 106
Physical Common Packet Channel 58
Physical Downlink Shared Channel 60
Physical Random Access Channel 58
PICH 62
pilot bits 52
PRACH 58
Preamble Initial Power 116
Primary Common Control Physical Channel 62
Primary Common Pilot Channel 62

Primary Synchronization Channel 62
Primary Synchronization Code 62
protocol stack 10
Protocol Stack on Iur-Interface 22
PSC 62
P-SCH 62
pseudo-noise code 28, 30
PSTN 6

Q

QE 98, 224
QPSK 28
Quality Estimate 98
Quality Estimate 224

R

Radio Link Control 118
Radio Link Protocol 14
Radio Network Controller 6
Radio Network Subsystem 6
Radio Resource Control 152
random access 102
random number 108
Reset Sequence Number 148
RLC-Size 122
RLIST-SUFI 144, 146
RLP 14
RNC 6
RNS 6
RNSAP 22, 360, 364
RRC 152
RRC-Idle Mode 160

UMTS – Signaling & Protocol Analysis

RRC-States	158
RSCP	222, 348
RSN	148
RSSI	222

S

SABP	20
SAPI	156
scaling factors	106
SCCP	16
S-CCPCH	60
S-CPICH	62
scrambling code	28, 30
SCTP	16
Secondary Common Control Physical Channel	60
Secondary Common Pilot Channel	62
Secondary Synchronization Channel	62
semi-static part	40
sequence number	124, 136
Served User Generator Reference	396
Serving GPRS Support Node	6
Serving RNC	8, 22
S-field	52
SGSN	6
Short Message Service Center	6
Signaling Connection Control Protocol	16
signaling radio bearer	156
size of a MAC-PDU	80
SM-SC	6
SN	124, 136
soft handover	98, 308, 320
softer handover	308, 312
split/combine-function	22, 98, 312
spreading code	28
spreading factor	58
SRB	156
SRNC	8
SRNC-Id	360, 364
SRNS-relocation	310
S-RNTI	86
S-SCH	62
SSDT	52
Stream Control Transfer Protocol	16
SUFI	140
Super Field	140
synchronization procedure B	312, 320
SYS_INFO2quater	348

T

T305	358
T307	358
T308	206
T317	358
Tag, Length, Value	174
Tag-Field	174
Target Channel Type Field	80
TB-size	122
TB-Size	40
TBS-Size	40
TCTF	80
TF	40
TFCI	48
TFCI-field	52
TFI	40, 42, 46
TFS	40
Timer_Discard (RLC)	150

© INACON GmbH 2003 - 2005. All rights reserved. Reproduction and/or unauthorized use of this material is prohibited and will be prosecuted to the full extent of German and international laws. Version Number: 1.6

UMTS – Signaling & Protocol Analysis

TLV-encoding	174
TM	118
TMD-PDU	80, 124
TPC-field	52
traffic channel	36
traffic volume measurements	100
transmission time interval	40
Transparent Mode	118
transport block size	122
transport channels	38
transport channels (common)	38
transport channels (dedicated)	38
transport format	40
Transport Format Combination	48
Transport Format Combination Identifier	48
Transport Format Indicator	40
transport format set	40
Transport Network Control Plane	24
TTI	40

U

UARFCN	26
UM	118
UMD-PDU	124
UMSC	6
Unacknowledged Mode	118
URA_PCH-State	168
U-RNTI	86
UTRA Carrier RSSI	222
UTRAN	6
Uu-interface	11, 15, 17

V

Visitor Location Register	6
VLR	6

W

WINDOW-SUFI	140